World Military Leaders

World Military LEADERS

A Collective and Comparative Analysis

MOSTAFA REJAI
AND KAY PHILLIPS

Westport, Connecticut
London

Library of Congress Cataloging-in-Publication Data

Rejai, M. (Mostafa)
    World military leaders : a collective and comparative analysis /
Mostafa Rejai and Kay Phillips.
    p. cm.
    Includes bibliographical references and index.
    ISBN 0–275–95386–6 (alk. paper)
    1. Command of troops. 2. Military biography. 3. Sociology,
Military. 4. Psychology, Military. I. Phillips, Kay. II. Title.
UB210.R45   1996
355.3'3—dc20   95–22011

British Library Cataloguing in Publication Data is available.

Copyright © 1996 by Mostafa Rejai and Kay Phillips

All rights reserved. No portion of this book may be
reproduced, by any process or technique, without the
express written consent of the publisher.

Library of Congress Catalog Card Number: 95–22011
ISBN: 0–275–95386–6

First published in 1996

Praeger Publishers, 88 Post Road West, Westport, CT 06881
An imprint of Greenwood Publishing Group, Inc.

Printed in the United States of America

The paper used in this book complies with the
Permanent Paper Standard issued by the National
Information Standards Organization (Z39.48–1984).

10 9 8 7 6 5 4 3 2 1

In Memoriam

JAMES SMOOT COLEMAN

(University of California, Los Angeles)

Contents

Tables	ix
Preface	xi
Acknowledgments	xiii
Introduction	xv

I. THEORY

1. Military Leaders: An Interactional Theory	3

II. FOUNDATIONS

2. General Patterns	13
3. Seeking Comparative Patterns	31

III. PROPELLANTS

Introduction	37
4. American Leaders	41
5. British Leaders	65
6. French Leaders	77
7. German Leaders	89
8. Other Leaders	97

IV. SYNTHESIS

9. Military Leaders in Theoretical Perspective   123

   Appendix: General Code Sheet   137
   Bibliography   149
   Index   161

Tables

2.1	Frequency Distribution of Hypothesized Characteristics: Age and Birthplace	15
2.2	Frequency Distribution of Hypothesized Characteristics: Socioeconomic Status and Family Life	17
2.3	Frequency Distribution of Hypothesized Characteristics: Ethnicity and Religion	20
2.4	Frequency Distribution of Hypothesized Characteristics: Education and Occupation	22
2.5	Frequency Distribution of Hypothesized Characteristics: Activities and Experiences Before Highest Rank	24
2.6	Frequency Distribution of Hypothesized Characteristics: Ideologies and Attitudes	27
9.1	Psychological Attributes of Military Leaders	126

Preface

Who are military leaders and what propels them toward the military profession? These are the dual questions to which this book seeks answers.

Focusing on forty-five major military figures from four continents and thirteen countries, spread across four centuries, this work presents, for the first time, a comparative, collective portrait of high-ranking military officers.

We begin by developing an interactional theory of military leaders, stressing the interplay between sociodemographic variables, psychological dynamics, and situational factors.

Focusing on sociodemographic traits, we examine age and birthplace, socioeconomic status, family life, ethnicity and religion, education and occupation, activities and experiences, and ideologies and attitudes. With some exceptions, we find military leaders to be a remarkably coherent and homogeneous group.

Turning to psychological dynamics, we find the leaders to be propelled toward military careers by some combination of nationalism, imperialism, relative deprivation, love deprivation, marginality, and vanity. (These concepts are defined in the Introduction to Part III.)

Among the situational variables, particularly important are birthplace, family influences, national crisis or emergency, and luck or chance.

We conclude by synthesizing our findings, offering some probabilistic statements concerning the major variables that may dispose a person toward a military career, and presenting some closing remarks.

Acknowledgments

In preparing the present study, we have had the assistance and cooperation of many individuals and institutions. We are happy to have the opportunity to acknowledge our debts.

Important research assistance was provided by Kathy Baum, Grant Benson, Caroline Berry, Arthur Chenin, Benjamin Griffiths, Craig Ilgenfritz, Robert Johnson, William Johnson, Tamara Jones, Kitty Kincaid, Mack Leftwich, Ronald O'Leary, and Lisa Schworm. Susan Bedel helped with the translation of Russian materials, and Robert McFadden with the translation of German sources.

Among Miami University librarians particularly resourceful at locating research materials, we note Sarah Barr, Penny Beile, Jennifer Block, John Diller, Rebecca Morgenson, Jenny Presnell, and Scott Van Dam.

For a third time, Shahla Mehdizadeh of Academic Computer Services, Miami University, generously shared with us her superb data analytical skills. John H. Skillings of the Department of Mathematics and Statistics proved to be an extremely knowledgeable consultant.

Some of the research for this study was conducted at the Center for International Affairs, Harvard University, and the Hoover Institution, Stanford University. We are grateful to the officers and staffs of these organizations for superbly productive opportunities.

Miami University has supported our work in every conceivable way through the years. Our colleague Warren L. Mason read the manuscript and offered a perceptive and constructive critique. Betty Marak "processed" successive drafts of the manuscript with unfailing expertise and good humor.

Dan Eades of Praeger Publishers remains a continuing source of inspiration and sound counsel. Sue Anderson did a superb job of copyediting. John Donohue of Rainsford Type and Ann Smith of Praeger skillfully steered the manuscript through production.

We alone remain responsible for any errors of fact or interpretation in connection with this work.

Introduction

This is a book about military leaders, not about military leadership.

We might note in passing, however, that the study of military leadership (as indeed *all* types of leadership) has not progressed much through the centuries. Although a great deal has been written about the subject, from Sun Tzu (fourth century B.C.) to Machiavelli (1521) to Clausewitz (1832) to de Gaulle (1960) to Montgomery (1961) to Keegan (1978, 1988), military leadership has been treated in terms of such concepts as character, courage, competence, intelligence, persuasiveness, decisiveness, discipline, mastery of the subject, commitment to high ideals, optimism, self-confidence, self-control, selflessness, concern for subordinates, luck, and the like (see also Beaumont 1974; Blumenson and Stokesbury 1975; Buck and Korb 1981; Carter 1952; Holmes 1989; Kellett 1982; Laffin 1966; Meier 1943; Miller 1920; Puryear 1971; Taylor and Rosenbach 1984; Wood 1984). While the foregoing authors have made important contributions, no unified theory of military leadership (or, for that matter, *any* type of leadership) has yet emerged.

On a related matter, although much has been written about *individual* military leaders, *comparative* portraits of high-ranking officers are extremely rare. An exception is Vagts (1937, 61–68), who offered some brief and general observations about the social composition of the English, French, and Prussian officers in the seventeenth, eighteenth, and nineteenth centuries. A more extensive (but narrower) study is Janowitz (1960, 1971), who focused on the social background, education, career patterns, and ideological orientations of a group of American military leaders between 1910 and 1950. An even narrower (and informal) work is Mylander (1974), who concentrated on American Army officers, emphasizing such topics as social background, education, career path, professional life, and retirement.

Focusing on the comparative, collective portrait, in this work we raise two interrelated sets of questions about military leaders—one dealing with their sociodemographic characteristics and traits, and the other with their motivation and emergence upon the scene.

1. Who are the men[1] who assume positions of military leadership? Are they urban dwellers or do they come from rural locations? What level of education do they have? What socioeconomic backgrounds do they represent? What sorts of family lives do they have? What ethnic and religious groups do they come from? Where and how do they acquire social awareness and military experience? What ideological postures do they adopt? What attitudes do they hold about man and society? Are there sociodemographic features that are "situational" in nature (that is, traits over which one has no personal control)?

2. Under what conditions do men seek positions of military leadership? What forces or circumstances provide impetus to the emergence of military elites? What socialization patterns do they encounter in childhood and youth? What are the forces or dynamics—social, psychological, situational—that catapult men into positions of military leadership?

Our selection of military leaders was guided by a series of explicit criteria. First, as our time referent we chose the period from the 1600s to the 1980s, thus maintaining consistency and comparability with our earlier studies of leaders (see Rejai and Phillips 1979, 1983, 1988). Second, we selected figures whose status as military leaders was confirmed through listings in major "who's who"-type sources (for instance, Dupuy, Johnson, and Bongard 1992; Keegan and Wheatcroft 1976; Martell and Hayes 1974; Spiller and Dawson 1989). Third, we excluded from consideration military leaders who were living as of the summer of 1988, when this project was launched (for example, James H. Doolittle, Saddam Hussein, Muammar el-Qaddafi, Itzhak Rabin, Matthew B. Ridgway, H. Norman Schwarzkopf, Ariel Sharon, William Westmoreland, and others). Fourth, other things being equal—and in the hope of facilitating data collection—we gave preference to the more famous military figures. Fifth, although there is an abundance of information about American military leaders, in the interest of balance we limited our selection to relatively few individuals. Sixth, in order to minimize sample contamination, we excluded from consideration such monarch-soldiers as England's Charles I (1600–1649), Prussia's Frederick the Great (1713–1768), Sweden's Gustavus Adolphus (1594–1633), Russia's Peter the Great (1672–1725), and several others. Seventh, we excluded from consideration military leaders we had studied in connection with our previous work (see Rejai and Phillips 1979, 1983, 1988, 1993). These include: Chiang Kai-shek (Jiang Ji-shi), Chou En-Lai (Zhou Enlai), Chu Teh (Zhu

De), Oliver Cromwell, Dwight D. Eisenhower, Vo Nguyên Giap, Ulysses S. Grant, Ernesto (Che) Guevara de la Serna, William H. Harrison, Andrew Jackson, Lin Piao (Lin Biao), Mao Tse-tung (Mao Zedong), Nicholas Nikolaievich, Rupert of the Rhine, Zachary Taylor, Leon Trotsky, and George Washington. From this list we have resurrected only three individuals for inclusion in the present work because, as we shall see, they provide special nuance. The three are Eisenhower, Jackson, and Washington.

We then began the tedious process of data collection on a uniform code sheet (see the Appendix). While data are abundant for such figures as Napoleon, Eisenhower, and de Gaulle, for example, in many cases, the sheer paucity of even plain sociodemographic data was so severe as to discourage even the most enthusiastic researcher. If sociodemographic information is so hard to come by, one can only imagine the state of social-psychological data, remembering that social-psychological studies of leaders constitute a relatively recent phenomenon.

Accordingly, we cut our list of military leaders to forty-five individuals for whom we had gathered fairly comprehensive data. The data, we should note, cover the period up to a leader's decision to join the military or of his reaching highest military rank; what happens to an individual thereafter does not concern us for the purposes of this work.

Organized by country, our list of forty-five military leaders is as follows.

America/U.S.A.
George Washington, 1732–1799

John Paul Jones (né John Paul, Jr.), 1747–1792

Andrew Jackson, 1767–1845

Robert Edward Lee, 1807–1870

William Tecumseh Sherman, 1820–1891

Alfred Thayer Mahan, 1840–1914

John Joseph Pershing, 1860–1948

George Catlett Marshall, Jr., 1880–1951

Douglas MacArthur, 1880–1964

William Frederick Halsey, Jr., 1882–1959

George Smith Patton, 1885–1945

Chester William Nimitz, 1885–1966

Claire Lee Chennault, 1893–1958

Dwight David Eisenhower, 1890–1969

Britain
Thomas Fairfax, Fairfax of Cameron, 3rd Baron, 1612–1671

John Churchill, 1st Duke of Marlborough, 1650–1722
John Burgoyne, 1722–1792
Horatio Nelson, 1st Viscount Nelson and Duke of Bronte, 1758–1805
Arthur Wellesley (né Wesley), 1st Duke of Wellington, 1769–1852
Charles George Gordon, 1833–1885
Horatio Herbert Kitchener, 1850–1916
Bernard Law Montgomery, 1st Viscount, Montgomery of Alamein, 1887–1976
Louis Francis Albert Victor Nicholas Mountbatten (né Battenberg), Mountbatten of Burma, 1900–1979

France
Napoleon Bonaparte (né Napoleone di Buonaparte), 1769–1821
Marie Joseph Paul Yves Roch Gilbert du Motier Lafayette, Marquis de, 1757–1834
Ferdinand Foch, 1851–1929
Henri Philippe Benoni Omer Joseph Pétain, 1856–1951
Charles André Joseph Marie de Gaulle, 1890–1970

Germany/Prussia
Carl von Clausewitz, 1780–1831
Helmuth Karl Bernhard Graf von Moltke, 1800–1891
Paul Ludwig Hans Anton von Beneckendorf und von Hindenburg, 1847–1934
Erich Friedrich Wilhelm von Ludendorff, 1865–1937
Erwin Johannes Eugen Rommel, 1891–1944

Israel/Palestine
Moshe Dayan, 1915–1981

Italy
Giuseppe Garibaldi, 1807–1882

Japan
Isoroku Yamamoto (né Takano), 1884–1943
Hideki Tojo, 1884–1948

Mexico
Antonio López de Santa Anna Perez de Lebron, 1794–1876

Russia/U.S.S.R.
Grigory Alexandrovich Potemkin, 1739–1791
Anton Ivanovich Denikin, 1872–1947
Georgi Konstantinovich Zhukov, 1896–1974

Spain
Francisco Franco y Bahamonde, 1892–1975

Turkey

Mustafa Kemal Atatürk (né Mustafa), 1881–1938

Venezuela

Simón José Antonio de la Trinidad Bolívar y Palacios, 1783–1830

Yugoslavia

Josip Broz Tito (né Josip Broz), 1892–1980

This, we should note, is the only place in the book where all the names appear in full. In the balance of the text, shorter or more familiar names will be used.

As can be seen, these military leaders come from four continents and thirteen countries, spread across four centuries. Moreover, they are somewhat evenly divided between twentieth-century leaders (25) and those who came in earlier centuries (20). They are also relatively evenly divided between underdeveloped (14), semideveloped (14), and developed (17) countries.

In selecting the fourteen American leaders, we faced a serious tradeoff: enhancing sample size had to be balanced against the introduction of possible cultural bias. However, even if we had reduced the number of Americans by, say, one-third or one-half, the sample as a whole would still retain a distinct Western bias. Therefore, we opted for enhancing sample size.

(*Note:* Although he was born in Scotland, we grouped John Paul Jones with the American leaders because he is routinely identified as an American naval hero. Although at times he considered himself a Citizen of the World, America remained his singular and enduring fascination. See Chapter 4.)

Regrettably, due to wide gaps in data, we were unable to include some very notable military leaders in our study. These include:

Mikhail Vasilyevich Alekseev, 1857–1918, Russia

Charles Cornwallis, 1738–1805, Britain

Joseph Jacques Césaire Joffre, 1852–1931, France

Antoine Henri de Jomini, 1779–1869, Switzerland

Lavr Georgievich Kornilov, 1870–1918, Russia

Alfred Graf von Schlieffen, 1833–1913, Germany

Henri de la Tour d'Auvergne Turenne, 1611–1675, France

It goes without saying that we do not have a representative sample of military leaders. In addition to being small, in the technical language of contemporary social science, ours is a nonrandom, purposive sample of military elites with possibilities as well as limitations. We shall return to this topic in Chapter 9.

The gathering of social background data on military leaders went hand in hand with the examination of the historical, situational, social, and psychological conditions under which they emerged. These endeavors involved close scrutiny of primary and secondary sources with heavy reliance on biographical and autobiographical materials, to the extent that these were available. Aware of the biases that may be embedded in such literature, both positive and negative, we treated the matter with due care and referred to as many independent sources as possible. (Data collection, we should note, was concluded in the summer of 1993.)

The questions of validity and reliability of the data collected for this work have been given careful consideration. As for validity, the data items used are adapted from the general literature of elite analysis; in fact, these items are by now part of the conventional wisdom, representing considerable scholarly consensus. Moreover, the variables for which data are collected interrelate to a large extent. Thus, demographic, ideological, attitudinal, social, psychological, and situational variables not only cohere as distinct groups, they intersect at many points as well.

As far as reliability is concerned, every effort has been made to minimize possible errors. While there is probably no such thing as an error-free study, a series of precautions was taken in data collection and coding. First, as the Bibliography indicates, a wide range of data sources was employed. Second, while the initial data gatherings were in the hands of the research assistants listed in the Acknowledgments, each assistant was provided with a detailed code sheet (see the Appendix) and with detailed instructions. Third, we jointly checked every coded item against the many sources indicated, thus arriving at penultimate coding decisions, while the second-named author assumed responsibility for all the final codings. This three-stage coding process assured the reliability of the data.

In all instances, problems of conflicting data were resolved by relying on two yardsticks: (1) the frequency with which a datum appeared in independent sources and (2) the authenticity of the source or person giving the datum. Needless to say, the second yardstick was given preference throughout.

As for the actual coding, most of the socioeconomic variables were relatively straightforward and presented no particular problems. Such matters as age, birthplace, socioeconomic status, ethnicity and religion, number of siblings and age ranking among them, education and occupation, father's education and occupation, travels, publications, and the like were relatively simple to isolate and code. On the other hand, such issues as home and school influences, attitude toward human beings ("nature of man"), attitude toward one's own country, and attitude toward the international community were based on close textual examination of the available source materials, which at times involved judgment calls. As may be expected, coding psychological variables proved the most problematical, a subject to which we return in the Introduction to Part III.

We have made every effort to keep the organization and presentation of our materials as straightforward as possible. Part I sets forth the theoretical foundations of the enterprise. Chapter 1 presents an interactional theory of military leaders in terms of the interplay of sociodemographic variables, psychological dynamics, and situational factors. It also presents a series of interrelated propositions and hypotheses derived from the interactional theory as well as from relevant research findings.

Part II undertakes a quantitative comparative analysis of military leaders in the light of the propositions and hypotheses developed in Chapter 1. Chapter 2 presents a collective portrait of the forty-five leaders in terms of a set of sociodemographic, experiential, ideological, and attitudinal variables. Chapter 3 employs factor analysis and discriminant analysis in unsuccessful efforts to compare the military leaders with two groups of loyalist and revolutionary elites we had studied in our previous work.

Part III undertakes a qualitative analysis of the social, psychological, and situational dynamics underlying the emergence of military leaders. Chapter 4 focuses on American leaders; Chapter 5, on the British; Chapter 6, on the French; Chapter 7, on the Germans; and Chapter 8, on the remaining leaders.

Part IV (Chapter 9) synthesizes the study by reviewing and consolidating the empirical findings, offering probabilistic statements to account for the emergence of military leaders, and suggesting some concluding thoughts.

Although we have endeavored to steer clear of technical jargon throughout our presentations, some use of cant has been unavoidable, particularly in Chapter 3 (which happens to be blissfully short). The uninterested reader may skip the affected passages without appreciable loss of content.

NOTE

1. Needless to say, as of this writing, women have not had an opportunity to emerge as military leaders.

PART I

Theory

Chapter 1

Military Leaders: An Interactional Theory

As noted in the Introduction, the subject of military leadership has commanded singular scholarly attention throughout history. Yet, as of this writing no satisfactory theory has emerged. In this book, we shift the focus of analysis from "leadership" to "leaders," and we examine their sociodemographic backgrounds and their emergence upon the scene. We label our effort an interactional theory of military leaders. This theory is a modified statement of a conceptual scheme we have used in our previous studies of loyalist and revolutionary political elites (see Rejai and Phillips 1979, 1983, 1988).

Recurrent in the literature of military leadership (indeed, of *all* leadership) is the concept of *situation*, variously defined. Recurrent also is the idea of leader personality or *psychology*. Moreover, we side with scholars who approach personality or psychology in such a comprehensive fashion as to incorporate leadership skills (in pursuit of goals and objectives). These skills, needless to say, are verbal, organizational, or (most likely) both. While we take for granted the centrality of a third variable—leader-follower interaction—we have no way of directly examining or demonstrating this relationship retroactively in the historical contexts with which we deal.

In the pages that follow, we shall define situation and psychology in explicit terms, and we shall apply our interactional perspective to a group of military leaders. Throughout, we shall raise such questions as: What kinds of sociodemographic traits do military leaders possess and how do these traits impinge on situations? What kinds of psychological forces characterize or influence military elites? What kinds of situational variables impinge on psychologies?

Consistent with our interactional stance, we shall present a series of hypotheses and propositions concerning military leaders. These propositions are drawn, with appropriate modifications, from our previous studies of leaders (see Rejai and Phillips, 1979, 1983, 1988, and the sources cited therein).

Support for our interactional theory may be garnered from talmudic times to the present. According to the Talmud (as quoted by Supreme Court Justice Sandra Day O'Connor [1982]), "In every age, there comes a time when leadership suddenly comes forth to meet the needs of the hour. And so there is no man [leader?] who does not find his time, and there is no hour that does not have its leader."

In more recent times, Barbara Tuchman has observed:

> The human being—you, I, or Napoleon—is unreliable as a scientific factor. In combination of personality, circumstance, and historical moment, each man is a package of variables impossible to duplicate. His birth, his parents, his siblings, his food, his home, his school, his economic and social status, his first job, his first girl, and the variables inherent in all of these, make up that mysterious compendium, personality—which combines with another set of variables: country, climate, time, and historical circumstance. (quoted in DeLuca 1983, 11)

An elaborate literature in the field of social psychology has been devoted to similar concerns. A recent contributor (Magnusson 1981) organized an entire book, *Toward a Psychology of Situations: An Interactional Perspective,* around three themes: (1) actual or objective environments and situations, (2) perceived or subjective environments and situations, and (3) person/situation interaction.

Consistent with our central concerns (see the Introduction), our theoretical position will be stated in terms of two interrelated sets of propositions and hypotheses about military leaders: What socioeconomic characteristics and traits do they have? How and why do they arise upon the scene?

CHARACTERISTICS OF MILITARY LEADERS

As noted in the Introduction, we study the social demographics of military leaders in order to isolate social background variables that may be "situational" in nature, thereby contributing to our theory of the emergence of military elites.

We expect military leadership to be positively correlated with middle age. Military leaders will be in their forties and fifties upon reaching their highest ranks. Although military elites may have been exposed to military ideologies—and may have participated in military activities—at much younger ages, the development of the requisite skills and the assumption of leadership roles are likely to require gestation and refinement. There may be instances, however, when crisis or chance play more compelling roles than one might logically anticipate.

Military leadership, we expect, is positively correlated with the heightened awareness and activism nurtured by urban life. Accordingly, military leaders either come from urban centers or, if born and raised in rural environments, they acquire early and sustained exposure to urban cultures. Early involvement in national affairs in urban areas—and the values and skills found therein—is likely to be pivotal to the development of military elites.

We expect military leadership to correlate positively with the dominant indigenous culture of a society. Military elites of all persuasions most likely belong to the main ethnic groups in their societies. Their religious backgrounds and orientations are also of the mainstream variety.

We project military elites to be somewhat representative of their respective social strata. We anticipate, however, that middle and upper classes will be overrepresented.

Should the foregoing propositions be confirmed by the data, it is logical to postulate that—given the high degree of predictability and continuity associated with their circumstances—military leaders are likely to experience stable and tranquil family lives. Instances of family conflicts and stormy childhoods will be rather rare.

As a component of their family lives, military leaders will have many siblings. However, consistent with the findings of our earlier studies of leaders, we expect the military men to be either the oldest or the youngest children. Middle children will be underrepresented among military elites because middle children are relatively passive and inactive.

Given their urban, social, and cultural backgrounds, we anticipate military leaders to be highly educated, many at prestigious military academies. It is a rare military leader who lacks a formal military education.

It follows that military leaders make the military their sole occupation; with rare exceptions (for instance, de Gaulle), they show little or no interest in politics. Upon retirement, however, some military leaders express interest in—or are recruited into—high political office or other occupational pursuits.

Consistent with their social backgrounds, we expect the occupations of *fathers* of military leaders to be of the prestigious variety. The military, the civil service, landed gentry, and the professions readily come to mind.

Consistent with their social backgrounds, military leaders are likely to be cosmopolitan in orientation. They travel far and wide, gaining exposure to various societies, languages, and cultures. Some military leaders tend to be prolific writers, concentrating chiefly (though not exclusively) on matters of military theory and practice.

Being regime-supportive, military leaders are likely either to maintain distance from politics or to participate only in legal political activities. As a rule, they adopt conservative and indigenous political ideologies. And, as a rule, they are free of harassment by the political authorities.

Given their ideological orientations, military leaders will tend to have a negative or pessimistic view of human nature. They will have uniformly positive attitudes toward their own countries, though their view of international society will be dualistic, identifying friends and enemies.

Given their high socioeconomic status, family background, and occupation, military leaders are strategically situated in their societies, commanding ready access to positions of power and authority.

EMERGENCE OF MILITARY LEADERS

The social backgrounds and socialization experiences of military leaders imbue them with sets of norms and values and prepare them for the roles they will be called upon to play in the future. By the same token, military leaders are in a position to internalize, articulate, and respond to the needs, desires, and aspirations of their subordinates. Failure to maintain vital ties with subordinates will impede or block reaching high office or, having reached it, such failure will spell demise.

But on a more basic level, what motivates a person to become a military leader? This deceptively simple question continues to elude theorists and researchers everywhere. Although the subject has received extensive attention in the social and behavioral sciences (for a literature review, see Rejai and Phillips 1988), in military science it has been largely ignored. Indeed, we have come across very few military scholars who have addressed the matter.

Witty and Lehman (1932) identified "nervous instability" as a possible source of military and political leadership, applying this perspective to Cromwell, Lincoln, Julius Caesar, Frederick the Great, and Napoleon. For instance, the authors noted Cromwell's religious devotion as a means of explaining uncommon cruelty, and they observed Napoleon's deep inferiority complex vis-à-vis the French.

Janowitz (1960, 1971) identified the following motive patterns for a group of American military leaders active between 1910 and 1950: family tradition, desire for social advancement, careerism, and code of honor.

Mylander (1974) concentrated on American Army officers, identifying ambition as the primary motivating force.

Dixon (1976) associated military competence with authoritarian leadership. He also noted the need for approval, code of honor, antiintellectualism, and machismo complex.

Keegan (1978) stressed the concept of honor in military leadership. And Keegan (1988) stressed the role of heroism.

Holmes (1989) focused on several factors in military leadership: personal honor or glory; glory of homeland, religion, or ideology; fear of cowardice and ridicule; self-sacrifice; and the need to eliminate a dehumanized enemy (as in the My Lai massacre of 1968).

Relying on this meager literature—and on our own previous studies of loyalist and revolutionary leaders—we shall focus on military motivation in terms of two indispensable concepts: psychology and situation.

The mental set or psychology to which we refer has five identifiable components. The hypotheses and propositions that follow are derived from the general literature of leadership motivation mentioned above and from our own previous studies of loyalist and revolutionary leaders (see Rejai and Phillips 1979, 1983, 1988).

Military elites are likely to be motivated by varieties of nationalism and patriotism. They seek to maintain the identity and integrity of their homelands. They set out to free their nations from the oppression and exploitation of other countries or, having reached this state, they may seek to expand their national frontiers. "Duty, Honor, Country"—need one say more?

Military elites are likely to be vain and egotistical in nature. Some are likely to be overly ambitious, exhibit heightened aggression, and be given to careerism. A degree of vanity appears to be an indispensable condition of *any* type of leadership role.

Some military leaders are driven by a compulsion to excel, to prove themselves, to overcompensate. This compulsion is likely due to feelings of inferiority, low self-esteem, or marginality arising from deviation from societal norms. How this condition comes about requires examination on a case-by-case basis.

Relative deprivation may serve as motive force of military orientation. Where there is a felt discrepancy between aspiration and achievement, one may redress the situation by gravitating toward the military profession. A military career affords a readily recognized road to social advancement.

Finally and relatedly, in a study of twenty-four British prime ministers—from Spencer Perceval in 1809 to Neville Chamberlain who resigned in 1940—Lucille Iremonger (1970) found that fifteen of them (62.5 percent) had lost one or both parents before reaching age fifteen. Finding this figure exceptionally high, she theorized that bereavement and the attendant love deprivation in childhood propelled the men in search of power and recognition in the political arena. We shall see whether Iremonger's theory applies to some military leaders as well.

In short, a variety of forces or dynamics play roles in shaping the mental set of military leaders. We hazard the proposition that no single motivation or dynamic is sufficient to explain the formation of all military personalities. Nor do we anticipate an invariant mix of psychological dynamics universally applicable to all military elites. A mix there shall be, to be sure, but we expect it to vary from leader to leader.

Though critical, psychology alone does not account for the emergence of military elites; it is a necessary but not a sufficient condition. For military leaders to emerge, it is imperative that psychology coincide with the presence of certain situations. Taken together, psychology and situation propel men toward the military profession.

Military leadership, we maintain, is largely situational in nature. Situations of crises—whether political, military, social, economic, or psychological—catapult military leaders into prominence and provide them with ready and willing subordinates. Political crises may consist of interelite rivalries or conflicts, riots or rebellions, mass violence and civil strife, or governmental corruption or ineptitude. Military crises are exemplified by war, coup d'état, or mutiny. Social crises include the disintegration of the prevailing ideology, normative order, and social institutions. Economic crises are represented by severe inflation or depression. Psychological crises may consist of relative deprivation, defined as perception of discrepancy between aspiration and achievement.

A second type of situation may account for the emergence of other military leaders. Specifically, the persistent turbulence and large-scale violence characteristic of the histories of Britain, France, Germany, Israel, Italy, Japan, Russia, the United States, and other countries facilitate the emergence of military leaders.

A third type of situation is to be found in the nationalist movements of the nineteenth and twentieth centuries, which sought to aggrandize various countries, pit the colonizer against the colonized, or both. Coming sooner or later, conflict is inherent in the very nature of nationalist movements. One thinks of America's experience with Native Americans, England, Mexico, Spain, the Philippines, and various Latin American countries. One is reminded of England's role in India and Southeast Asia. One thinks of France's experiences in Algeria and Indochina. One is reminded of Germany's confrontations with Europe. One thinks of Russian expansionism vis-à-vis the various "Republics" and the Baltic states. One is reminded of the tumultuous histories of Israel, Italy, Japan, Mexico, Turkey, Yugoslavia, and other countries.

A fourth type of situation is to be found in family tradition. Some military leaders are born to "military" families, being gradually and consistently socialized into a military culture.

A fifth type of situational dynamic is simply the role of chance, luck, or happenstance. Some military leaders gravitate toward the military not as a result of determination and planning but because of fortuitous circumstances or lacking a viable alternative.

Most of the foregoing types of situations operate at the societal or historical levels. A final set of situational variables relate to personal traits or attributes that are "external" to military leaders and over which they have, as individuals, no control. These include: birthplace, exposure to urban culture, socioeconomic status, family life (whether tranquil or stormy), number of siblings, age ranking among siblings, ethnicity, and religion.

It is clear that we use the idea of "situation" in four explicit and identifiable senses: (1) family tradition; (2) conditions of open or latent conflicts in which elements of power contests are sufficiently salient to be unavoidable; (3) conditions of luck, chance, or happenstance; and (4) conditions wherein certain personal attributes beyond one's sway set the stage for the assumption of leadership roles.

SUMMARY

Our interactional theory of military leaders is synthetic in that it integrates sociodemographic, psychological, and situational variables in the analysis of military personalities. The theory focuses on the salience of a series of social-background variables, and it stresses the interplay of psychology and situation in the emergence of military leaders. Taken together, these variables demonstrate why it is that: (1) not all situations give rise to military leaders, and (2) not all persons with the appropriate social-psychological traits emerge as military elites.

We now turn to an assessment of our theory against both quantitative (structured) and qualitative (unstructured) data.

PART II

Foundations

Chapter 2

General Patterns

In order to test the hypotheses and propositions advanced in Chapter 1 relative to the sociodemographic, experiential, ideological, and attitudinal attributes of military leaders, we gathered data from a variety of sources, as discussed in the Introduction and detailed in the Bibliography. In the pages that follow, we report the general results of our investigations.

In an effort to bring additional substance to our findings, we used four control variables: time period, level of development of the country, socioeconomic status (social class), and ideology. Only time period and social class yielded additional findings, and these findings are reported in tandem with the cross-tabulations, as appropriate. More precisely, time period revealed distinctions between twentieth-century leaders and those who came in the seventeenth, eighteenth, and nineteenth centuries. Similarly, even in societies separated by centuries and cultures, social class impinges on life chances and experiences.

Level of development did not yield additional findings, suggesting that leader attributes are highly similar in underdeveloped, semideveloped, and developed countries. Similarly, ideology did not prove helpful as a control variable because, as we shall see, for the most part military leaders adopt a relatively small number of conservative and moderate belief systems. Overall, as it will become clear, across time, space, or any other variable, military leaders represent a highly coherent and homogeneous group of men.

In the tabular presentations that follow, *totals do not always equal 45 because of missing data.*

FINDINGS

Age and Birthplace

Overwhelmingly, the military leaders in this study are older, mature men when they reach the highest rank. Table 2.1(A) shows that none achieved that rank before age twenty-five, with almost 70 percent reaching it after age forty-five. A significant difference occurs by time period. Prior to the twentieth century, over half (55 percent) of the leaders had reached the highest rank by age forty-five. During the twentieth century, only three men (12 percent) were awarded the highest rank by age forty-five: Dayan at thirty-four, Atatürk at forty, Denikin at forty-three. On the other hand, only four leaders assumed the highest rank beyond age sixty-five: Mahan at sixty-six, Foch at sixty-seven, Hindenburg at sixty-nine, Moltke at seventy-one. In addition to Dayan, the four youngest commanders are Bonaparte at twenty-six, Bolívar at thirty, Lafayette at thirty-two, and Fairfax at thirty-three. At thirty-four, Dayan is the youngest twentieth-century leader to ascend to generalship. Although by no means considered youngsters, within our sample of forty-five leaders, these five men represent a small group who ascended the ladder of command early. Perhaps the less complex technology of warfare and command prior to the twentieth century allowed these men to rise to early prominence. Needless to say, Dayan remains an exception. It is also noteworthy that Bolívar, Bonaparte, Dayan, and Lafayette rose to generalship in revolutionary situations.

Table 2.1(B) reveals that while exposed to military ideology early in life—with over 75 percent being younger than fifteen at exposure—a few came later to military ideology, and only one after age twenty-five (Garibaldi at twenty-six). Similarly, the leaders see military action or combat early in their careers: well over half (nearly 57 percent) are engaged in armed conflict prior to age twenty-five. Only three men saw actual combat past age forty-five: Ludendorff at forty-nine, Eisenhower at fifty-two, Moltke at sixty-four.

Contrary to our hypothesis in Chapter 1, Table 2.1(D) indicates that military leaders are primarily rural born (almost 65 percent). Consistent with our expectations, however, Table 2.1(E and F) document that *all* the nonurban leaders develop early and sustained exposure to urban cultures. Urban cultures—and the values and skills they afford—are pivotal in the emergence of all leaders, military and otherwise.

Socioeconomic Status and Family Life

The socioeconomic status of military leaders is not entirely consistent with our hypotheses presented earlier. Table 2.2(A) shows that half of

Table 2.1
Frequency Distribution of Hypothesized Characteristics: Age and Birthplace

(A) Age at Highest Rank: By Time Period

Age	Before 20th century N	%	20th century N	%	Total N	%
25-34	4	20.0	1	4.0	5	11.1
35-44	7	35.0	2	8.0	9	20.0
45-64	7	35.0	20	80.0	27	60.0
65+	2	10.0	2	8.0	4	8.9
Total	20	44.4	25	55.6	45	100.0

Gamma = .56   Chi square = 10.41 $p < .015$

(B) Age First Exposed to Military Ideology

Age	N	%
0-15	34	75.6
16-19	7	15.6
20-24	3	6.7
25-34	1	2.2
Total	45	100.0

(C) Age First Took Part in Military Action (Combat)

Age	N	%
0-15	3	6.7
16-19	8	17.8
20-24	14	31.1
25-34	15	33.3
35-44	2	4.4
45-64	3	6.7
Total	45	100.0

(D) Birthplace

Birthplace	N	%
Urban	16	35.6
Rural	29	64.4
Total	45	100.0

Table 2.1 (Continued)

(E) Exposure to Urban Life if Nonurban Born

Number of years of exposure	N	%
1-3	0	0.0
4 or more years	29	100.0
Total	29	100.0

(F) Age Exposed to Urban Life if Nonurban Born

Age	N	%
0-14	17	58.6
15-19	10	34.5
20-24	2	6.9
Total	29	100.0

them have middle class backgrounds while slightly less than 30 percent come from the upper class. The military profession continues to afford opportunities for upward social mobility, for 20 percent of the total group emerged from the lower class. With formal military training and a military subculture that instills socially appropriate behaviors, by the time a leader has reached the highest rank, he has acquired the trappings and the veneer of upper middle class status.

Further review of Table 2.2(A) allows us to note differences in the social background of the leaders by time period. Prior to the twentieth century, 50 percent of the leaders came from the upper class, 40 percent from the middle class, and only 10 percent from the lower class. (Only two early military leaders—Jackson and Garibaldi—have lower class origins.) In the twentieth century, however, a major shift in class origins occurs: only 12 percent from the upper class, 60 percent from the middle class, and 28 percent from the lower class. In other words, upper class representation has sharply dropped, while middle class representation has risen by 50 percent and lower class representation has nearly tripled. In the twentieth century, significant opportunities for upward social mobility are afforded to men with aspirations for military careers.

As Table 2.2(B) indicates, military leaders are overwhelmingly of le-

Table 2.2
Frequency Distribution of Hypothesized Characteristics: Socioeconomic Status and Family Life

(A) Socioeconomic Status: By Time Period

Status	Before 20th century N	%	20th century N	%	Total N	%
Upper class	10	50.0	3	12.0	13	28.9
Middle class	8	40.0	15	60.0	23	51.1
Lower class	2	10.0	7	28.0	9	20.0
Total	20	44.4	25	55.6	45	100.0

Gamma = .65   Chi square = 8.22 $p < .016$

(B) Legitimacy Status

Status	N	%
Legitimate	43	95.6
Illegitimate	1	2.2
Illegitimate, parents married	1	2.2
Total	45	100.0

(C) Number of Siblings

Number of siblings	N	%
None	3	6.7
One	6	13.3
Two	5	11.1
Three	5	11.1
Four	5	11.1
Five	6	13.3
Six	3	6.7
Seven or more	11	24.4
Total	44	100.0

Table 2.2 (Continued)

(D) Age Ranking Among Siblings

Age ranking	N	%
Only child	3	6.8
Youngest	6	13.6
Middle	20	45.5
Oldest	10	22.7
Oldest son	4	9.1
Other	1	2.3
Total	44	100.0

(E) Family Life Character: By Time Period

Character	Before 20th century		20th century		Total	
	N	%	N	%	N	%
Broken home	8	42.1	2	8.0	10	22.7
Tranquil	7	36.8	20	80.0	27	61.4
Stormy	3	15.8	3	12.0	6	13.6
Other	1	5.3	0	0.0	1	2.3
Total	19	43.2	25	56.8	44	100.0

Cramer's V = .48   Chi square = 10.23 $p < .017$

gitimate birth. Only Burgoyne is illegitimate, and Dayan's parents married upon learning of pregnancy (see Chapters 5 and 8).

Born to families following the normative order of their societies, military leaders tend to have many siblings. Even though over half of the leaders are from the twentieth century, Table 2.2(C) reveals that only nine are either only children or grow up with one sibling, while the other thirty-five have two or more siblings. (Sibling data are missing for Santa Anna.) Nearly a quarter of the leaders have seven or more siblings. Tito tops the list with fourteen siblings, while Sherman has sixteen natural and adopted siblings.

Table 2.2(D) unravels the age ranking of the military leaders among their siblings. As can be seen, middle children constitute 45.5 percent of the sample, while *only,* youngest, oldest, and oldest sons make up just over 50 percent. (The "Other" leader is Washington, who is the oldest son of a second marriage.) This finding somewhat departed from what we had discovered in a previous study of leaders (Rejai and Phillips 1988, 47–49), where middle children were decidedly underrepresented and other children decidedly overrepresented. In the earlier study, we characterized the oldest children as "Princes," the youngest children as "Conquerors," and the middle children as "Diplomats," suggesting that Princes and Conquerors play active leadership roles while Diplomats are relatively inactive, pragmatic, and conforming. While there are definitely Princes and Conquerors among our military leaders, the proportion of Diplomats is relatively high. Accordingly, it may be that the ordered and predictable life of the military—particularly in peacetime—is attractive to certain personality types.

Table 2.2(E) documents that military leaders tend to grow up in relatively tranquil family circumstances, with over 20 percent coming from broken homes (due to death of a parent). Controlling for time period shows that prior to the twentieth century, almost half (over 42 percent) of the leaders came from broken homes, while in the twentieth century only two leaders (Chennault and Franco) suffered similar fates. Overall only six leaders experienced stormy childhoods: Atatürk, Bolívar, Burgoyne, Dayan, Montgomery, and Potemkin. The relatively placid nature of many military men's home life may predispose them to the order and security that a peaceful military career affords.

Ethnicity and Religion

As hypothesized in Chapter 1, Table 2.3(A) points out that over four-fifths of the military leaders are from the main ethnic groups of their countries. Five leaders are from large ethnic minorities: Gordon (Scottish), Montgomery (Irish), Napoleon (Italian), Santa Anna (Creole), Wellesley (Irish). One person is from a small minority: Franco (Galician). The two individuals belonging to the "Other" category are Tito, who was a minority Croat before Yugoslav independence in 1918, and Dayan, who was a minority Jew before the founding of Israel in 1948. (We should note, however, that there was never a clear ethnic majority in the former Yugoslavia.)

Table 2.3(B) notes that military leaders come universally from the main religious groups of their respective societies. Table 2.3(C) shows a wide variety of religious *backgrounds*, from Christian to Jewish to Muslim to Buddhist. Table 2.3(D) indicates that four individuals changed their religious *orientations* by becoming atheists: Atatürk, Jones, Tito, Zhukov. The two persons in the "Other" category are: (1) Napoleon, who claimed he could be a Christian, a Jew, or a Muslim, and who for a time was a Jacobin and an atheist; and (2) Bolívar, who was an agnostic and indif-

Table 2.3
Frequency Distribution of Hypothesized Characteristics: Ethnicity and Religion

(A) Ethnicity

Ethnicity	N	%
Majority	37	82.2
Large minority	5	11.1
Small minority	1	2.2
Other	2	4.4
Total	45	100.0

(B) Religious Affiliation

Affiliation	N	%
Main group	45	100.0
Minority group	0	0.0
Total	45	100.0

(C) Religious Background

Background	N	%
Protestant	28	62.2
Catholic	13	28.9
Jewish	1	2.2
Muslim	1	2.2
Buddhist	2	4.4
Total	45	100.0

(D) Religious Orientation: By Social Class

Orientation	Upper class N	%	Middle class N	%	Lower class N	%	Total N	%
Atheist	0	0.0	0	0.0	4	44.4	4	8.9
Protestant	9	69.2	16	69.6	2	22.2	27	60.0
Catholic	3	23.1	4	17.4	2	22.2	9	20.0
Jewish	0	0.0	0	0.0	1	11.1	1	2.2
Buddhist	0	0.0	2	8.7	0	0.0	2	4.4
Other	1	7.7	1	4.3	0	0.0	2	4.4
Total	13	28.9	23	51.1	9	20.0	45	100.0

Cramer's V = .53   Chi square = 25.46 $p < .005$

ferent to religion, but was outwardly respectful of the Church because of the Church's influence throughout Latin America.

Education and Occupation

We expected military leaders to be relatively well educated, with many securing professional military training. Table 2.4(A) demonstrates the educational achievements of the forty-four leaders for whom we have data. (Data are missing for Santa Anna.) As can be seen, an overwhelming 75 percent (thirty-three) attended military academies, studying the art and science of warfare. Only two attended traditional liberal arts institutions, and nine (over 20 percent) either had no formal education or at most the equivalent of a high school degree.

As may be expected, leaders who achieved their highest rank prior to the twentieth century were much more likely to have limited educational opportunities, with over 40 percent having no formal education or a high school equivalency. However, over half of the early leaders (53 percent) were educated in military institutions. In the twentieth century, military leaders typically (92 percent) receive formal, professional education in military academies. Only Tito had very limited education, having attended a trade school and having worked as a locksmith for a time. But Tito is an exception in many ways (see Chapter 8).

Table 2.4(B) demonstrates the occupations of military leaders by social class. Given that social class impinges on opportunities for education and training, the differences revealed in this table are not surprising. While an upper class background accounts for nearly 54 percent of the individuals who attend military academies, another six upper class men combine the military profession with another occupation, typically statecraft or politics (for instance, Bolívar, Lafayette, and Mountbatten). A large majority of middle class leaders hold the military as their sole occupation. Only one military leader's occupation can be classified as professional revolutionary: Josip Tito, to whom we shall return in Chapter 8. As is apparent, once again, the military represents a path of social mobility for middle and lower class individuals.

Table 2.4(C) sheds further light on the social background of military leaders by focusing on the occupations of their fathers. As can be seen, over 35 percent of the fathers—the largest group—have a military occupation, over 13 percent come from the professions, nearly 18 percent have a working class background, another 18 percent hold a combination of occupations, only one father is a government official, and three fathers come from the field of business. Two factors stand out in this table: (1) the career paths of military leaders are significantly influenced by the military careers of their fathers; (2) once again, the military emerges as

Table 2.4
Frequency Distribution of Hypothesized Characteristics: Education and Occupation

(A) Highest Education Level Attained: By Time Period

Education level	Before 20th century N	%	20th century N	%	Total N	%
None, through high school	8	42.1	1	4.0	9	20.5
Some college, to B.A.	1	5.3	1	4.0	2	4.5
Military	10	52.6	23	92.0	33	75.0
Total	19	43.2	25	56.8	44	100.0

Cramer's V = .48   Chi square = 9.93 $p < .007$

(B) Primary Occupation: By Social Class

Occupation	Upper class N	%	Middle class N	%	Lower class N	%	Total N	%
Military	7	53.8	20	87.0	7	77.8	34	75.6
Professional revolutionary	0	0.0	0	0.0	1	11.1	1	2.2
Combination	6	46.2	3	13.0	1	11.1	10	22.2
Total	13	28.9	23	51.1	9	20.0	45	100.0

Cramer's V = .33   Chi square = 9.93 $p < .04$

(C) Father's Primary Occupation

Occupation	N	%
Military	16	35.6
Professions	6	13.3
Working class	8	17.8
Combination	8	17.8
Government official	1	2.2
Business	3	6.7
Landlord	3	6.7
Total	45	100.0

Activities and Experiences

Military leaders engage in a variety of activities and experiences, as exhibited in Table 2.5.

Military leaders publish treatises, typically on matters of military theory and practice—and occasionally on politics, philosophy, and literature as well. Table 2.5(A) reveals that only twelve of the forty-five leaders have no publications, while the other thirty-three have few, some, or many writings to their credit. The fourteen (over 30 percent) most prolific contributors are Atatürk, Bolívar, Bonaparte, Chennault, Clausewitz, de Gaulle, Denikin, Jones, Mahan, MacArthur, Montgomery, Nimitz, Pershing, and Santa Anna.

Table 2.5(B) suggests that military leaders are overwhelmingly involved in legal political organizations, while Table 2.5(C) documents that twelve of them (27 percent) participate in revolutionary organizations as well. The latter group includes Atatürk, Bonaparte, Dayan, Garibaldi, Jackson, Lafayette, Lee, Potemkin, Santa Anna, Tito, Washington, and Zhukov.

Given their involvement in revolutionary organization and activity, it follows that some military leaders compile records of arrest, imprisonment, or exile. Table 2.5(D) documents that while over three-quarters (77 percent) of leaders from the upper class have never been arrested, imprisoned, or exiled, over three-quarters (78 percent) of individuals from the lower class have been so detained. Table 2.5(E) points out that three leaders from the upper class, four from the middle class, and seven from the lower class spent time in prison or exile.

Military leaders are not as cosmopolitan as we had anticipated. Table 2.5(F) points out that over one-third of the leaders speak no language but their own, while another one-third speak one other language and three individuals are bilingual. Table 2.5(G, H, and I) note that military leaders have very limited experience traveling to other countries, and that fully 40 percent never left the borders of their own lands before reaching their highest ranks. (We discount as foreign travel military missions abroad or military incursions into other lands.) As a result, as Table 2.5(J) shows, they cultivate few or no foreign contacts.

Ideologies and Attitudes

The parochialism of military leaders is also reflected in the ideologies and attitudes they embrace. As Table 2.6(A) indicates, thirty-nine leaders (87 percent) accept an indigenous ideology, three hold foreign ideologies,

Table 2.5
Frequency Distribution of Hypothesized Characteristics: Activities and Experiences Before Highest Rank

(A) Publication Record

Publications	N	%
None	12	26.7
Few (1-3)	11	24.4
Some (4-6)	8	17.8
Many (7+)	14	31.1
Total	45	100.0

(B) Membership in Legal Political Organizations

Membership	N	%
No	6	14.0
Yes	37	86.0
Total	43	100.0

(C) Membership in Revolutionary Organizations

Membership	N	%
No	32	72.7
Yes	12	27.3
Total	44	100.0

(D) Arrest, Imprisonment, or Exile Record: By Social Class

Record	Upper class N	%	Middle class N	%	Lower class N	%	Total N	%
None	10	76.9	19	82.6	2	22.2	31	68.9
Some (1-3 times)	2	15.4	4	17.4	6	66.7	12	26.7
Moderate (4-6 times)	1	7.7	0	0.0	0	0.0	1	2.2
Frequent (7 or more times)	0	0.0	0	0.0	1	11.1	1	2.2
Total	13	28.9	23	51.1	9	20.0	45	100.0

Gamma = .51   Chi square = 16.82 $p < .01$

Table 2.5 (Continued)

(E) Duration of Imprisonment: By Social Class

Duration	Upper class N	%	Middle class N	%	Lower class N	%	Total N	%
None	10	76.9	19	82.6	2	22.2	31	68.9
1 year or less	0	0.0	1	4.3	2	22.2	3	6.7
2-9 years	2	15.4	3	13.0	4	44.4	9	20.0
10 or more years	1	7.7	0	0.0	1	11.1	2	4.4
Total	13	28.9	23	51.1	9	20.0	45	100.0

Gamma = .46   Chi square = 13.53 $p < .04$

(F) Foreign Languages

Number of languages	N	%
None	16	36.4
One	14	31.8
Two	6	13.6
Three	3	6.8
Four	1	2.3
Five	1	2.3
Bilingual	3	6.8
Total	44	100.0

(G) Foreign Travel Before Attaining Highest Rank

Travel	N	%
None	18	40.0
Little (1 country)	4	8.9
Moderate (2-3 countries)	15	33.3
Extensive (4 or more countries)	8	17.8
Total	45	100.0

Table 2.5 (Continued)

(H) Foreign Travel Before Attaining Highest Rank: Duration

Duration	N	%
None	18	40.0
Less than 1 year	12	26.7
1-3 years	6	13.3
4+ years	9	20.0
Total	45	100.0

(I) Foreign Travel Before Attaining Highest Rank: Place

Place	N	%
None	18	40.0
Europe	18	40.0
U.S.A.	1	2.2
Asia	1	2.2
Latin America	1	2.2
Combination	6	13.3
Total	45	100.0

(J) Continuing Foreign Contacts

Foreign contacts	N	%
None	27	77.1
Few (1-3)	3	8.6
Many (4 or more)	5	14.3
Total	35	100.0

and another three adapt a foreign ideology to the conditions of their own countries. Table 2.6(B) outlines the specific ideologies the leaders hold. While the spectrum is rather wide, the most common ideologies are conservative (or ultra-) nationalism (49 percent), democracy (27 percent), and rightist beliefs (11 percent).

As Table 2.6(C) demonstrates, most military leaders have heroes they respect and admire. These heroes are typically military figures from their

Table 2.6
Frequency Distribution of Hypothesized Characteristics: Ideologies and Attitudes

(A) Source of Ideology

Source	N	%
Indigenous	39	86.7
Foreign	3	6.7
Foreign/indigenous	3	6.7
Total	45	100.0

(B) Type of Ideology

Type	N	%
Democratic	12	26.7
Marxist-Leninist	1	2.2
Nationalist/Marxist-Leninist	1	2.2
Nationalist	1	2.2
Rightist	5	11.1
Conservative Nationalist	22	48.9
Vacillating	3	6.7
Total	45	100.0

(C) Admired Personalities

	N	%
Yes	36	80.0
Unknown	9	20.0
Total	45	100.0

(D) Attitude Toward Humans

Attitude	N	%
Negative	2	4.4
Dualistic	43	95.6
Total	45	100.0

Table 2.6 (Continued)

(E) Attitude Toward Own Country

Attitude	N	%
Fluctuating	1	2.2
Positive	44	97.8
Total	45	100.0

(F) Attitude Toward International Community

Attitude	N	%
Positive	1	2.2
Dualistic	44	97.8
Total	45	100.0

own countries or from other lands. The most admired figures are Alexander the Great, Julius Caesar, Napoleon, Nelson, and Washington. Some military leaders admire some of their immediate colleagues, while others admire members of their own families (typically a father or an older brother).

Table 2.6(D) reveals that virtually all military leaders have a dualistic view of human nature, distinguishing between the peoples of their own countries (= good) and other peoples (= evil). Only two leaders—Atatürk and Franco—have a universally pessimistic view of human nature, being deeply skeptical of the ability of their own peoples to rule themselves.

Table 2.6(E) suggests that, with a single exception, military leaders have positive views of their own countries. The sole exception is Napoleon, who fluctuates between being a passionate Corsican and a passionate Francophile. Similarly, as Table 2.6(F) points out, military leaders almost universally share a dualistic view of international society, distinguishing friends and enemies. The only exception is John Paul Jones, who has a positive view of the international community.

SUMMARY

With only three exceptions, the data support the attributes and characteristics of military leaders hypothesized in Chapter 1. Moreover, as anticipated in Chapter 1, some of our findings are "situational" in nature—that is, attributes over which the leaders have little or no control. These include: birthplace, ethnicity and religion, social class, father's occupation, early exposure to military ideology and to military action. These attributes lend additional credence to our interactional theory of military leaders.

As for the three exceptions: first, over time the social composition of military leaders has changed sharply in favor of the middle and lower classes, making the military increasingly an avenue of upward social mobility. Second, as far as sibling ranking is concerned, middle children are overrepresented among military elites while other children are underrepresented, challenging our earlier findings concerning the inactivism of middle children and suggesting that the secure life of the military—particularly in peacetime—may be attractive to certain personality types. Third, military leaders are not as cosmopolitan as we had anticipated.

These exceptions notwithstanding, military leaders constitute a remarkably homogeneous group. Given their occupation and status, they are strategically situated in their societies, commanding ready access to positions of power and authority.

Chapter 3

Seeking Comparative Patterns

Do military leaders represent a category of leadership which draws men with the particular characteristics, traits, and attitudes analyzed in the previous chapter? Are military leaders significantly different from leaders in other arenas of life and, if so, are they so different that they constitute a unique group? If so, what are the differences? Does leadership embrace such a diversity of attributes, traits, and experiences that no identifiable group readily emerges?

In a previous work (Rejai and Phillips 1988), we have dealt with two other groups of leaders: fifty loyalist leaders who sought to maintain the status quo, and fifty revolutionary leaders who set out to overthrow it. Leadership in general may be a phenomenon that requires and attracts individuals with identifiable traits and skills. Different *kinds* of leadership may be associated with those who possess or cultivate *different* traits and skills. Is it possible to compare and contrast the loyalists, the revolutionaries, and the generals?

Two methodological techniques are available for our purposes. Factor analysis of the demographic, social, and experiential traits of the three leadership groups may help us determine if leadership traits themselves can be identified and grouped. Discriminant analysis can help us determine if different types of leadership are associated with men whose characteristics are unique and isolable.

Accordingly, using as a base the three data sets collected over a period of two and a half decades, we first ran factor analyses of 144 leaders: 50 loyalists, 49 revolutionaries, and 45 generals. (We grouped George Washington with military leaders, even though previously we had studied him as a revolutionary, because his military career predated his revolutionary activities.) We then followed up the factor analyses with a discriminant analysis.

In determining which variables to use in the factor and discriminant analyses, we took care not to include a variable or trait that would readily identify a particular leadership group. For instance, in the range of possible occupations, we omitted that of professional revolutionary because it would be associated almost exclusively with one type of leader. For the same reason, we omitted the military occupation.

The list of variables that were included in the analyses, then, represents traits and experiences that are not restricted to a particular type of leader. These variables are: (1) type of country; (2) age exposed to political or military ideology, age at the time of highest office or of revolution or of highest rank, age first politically or militarily active; (3) family life, including siblings and age ranking among siblings; (4) socioeconomic status; (5) religious orientation; (6) education, including location (domestic or foreign), highest level attained, field; (7) primary occupation other than that as loyalist, revolutionary, or general; (8) father's primary occupation; (9) cosmopolitanism, including foreign languages, extent and duration of foreign travel, continuing foreign contacts; (10) publication record; and (11) attitude toward human beings and toward their own countries.

We addressed the problem of missing data by substituting modal values of the appropriate leadership group for nominal variables, and mean values of the appropriate leadership group for ordinal and interval variables. (For further information about our data categories, see Rejai and Phillips 1988, Appendix D, Code Sheet for Factor and Discriminant Analyses.)

FACTOR ANALYSIS

Factor analyses were performed on the set of sociodemographic, experiential, and attitudinal data for all the 144 leaders using SPSS^x (1988). R-factor analysis was executed in an effort to produce constellations of attributes the leaders may share; Q-factor analysis was intended to produce groupings of the leaders themselves.

While we recalled from earlier experience that factor analyses on the loyalists and revolutionaries had been inconclusive, we hoped that the addition of military leaders might change the outcome. Unhappily, it did not.

R-factor analysis did produce nine factors which explained over half of the variance in the data, but no conclusive set of traits emerged to shed further light on our investigation. For instance, the most significant factor explaining less than 15 percent of the variance could be labeled cosmopolitanism, but this finding did not take us much beyond the discussion in Chapter 2.

It appears that the loyalists, the revolutionaries, and the generals share such a wide constellation of sociodemographic, experiential, and attitudinal traits that no neat clusters of identifiable variables are forthcoming.

Having given up on the entire group of 144 leaders, we ran factor analyses on the forty-five generals by themselves. Not surprisingly, once again, our efforts were unsuccessful, as we came up with no appreciable (or reportable) findings.

DISCRIMINANT ANALYSIS

Discriminant analysis is a multivariate statistical technique designed with the explicit purpose of investigating differences between two or more groups of subjects on several dimensions or variables simultaneously (see Klecka 1980). Specifically, discriminant analysis is useful in studying the three leadership groups in two ways: (1) given a series of variables, it helps to predict to which group a particular leader belongs; and (2) it helps identify the discriminating power of each variable in determining group membership, thereby providing a ranking of the most important variables. In other words, if using sociodemographic, experiential, and attitudinal characteristics, discriminant analysis can predict to which group (loyalist, revolutionary, or military) a leader belongs, then those characteristics are in fact different for the three leadership groups. That being the case, it is more understandable that the factor analysis of the three groups combined did not yield productive results.

A stepwise discriminant analysis with the variables outlined above produced two discriminant functions, each composed of seventeen specific variables having the most discriminating power. The first discriminant function separated the military leaders from the revolutionaries; the second grouped the loyalists by themselves. While this comparison showed that military leaders are significantly different from loyalist or revolutionary leaders, discriminant analysis did not elucidate our subjects beyond the discussion in Chapter 2. Accordingly, we dispense with the presentation of our "findings."

SUMMARY

Lack of success with factor and discriminant analyses leads us to conclude that leadership per se cannot be identified exclusively by sociodemographic, experiential, and attitudinal data. As we shall see in Parts III and IV, the addition of motivational, psychological, and situational data is essential to rounding out our investigation.

PART III

Propellants

Introduction

According to our interactional theory (see Part I), military leaders emerge upon the scene in response to the interplay of three sets of forces: the sociodemographic, the psychological, and the situational. Having considered the first set of variables in Part II, we now turn to an examination of psychological and situational factors.

In various combinations, six psychological or motivational dynamics propel men toward military careers: (1) nationalism; (2) conservative (or ultra-) nationalism; (3) relative deprivation; (4) love deprivation; (5) marginality; and (6) vanity, egotism, narcissism. These psychological attributes, far from being new to our work, have been used frequently by a variety of writers and scholars in the analysis of major political, military, social, and religious figures (for the specifics, see Rejai and Phillips 1988, 71).

Coding psychological variables is among the most demanding tasks social scientists are called upon to undertake. The process is full of hazards and pitfalls, and capable of producing misleading results. The standard technique for the kind of psychological investigation at hand is content analysis. Under this method, the social scientist identifies representative passages from a leader's autobiographical writings or other pronouncements bearing on the topic(s) under study; then, under a rigorous set of controls, he or she subjects the writings to coding, quantification, analysis, and interpretation.

Its many problems notwithstanding, the method of content analysis is infeasible for the purposes of this work. The main reason is straightforward: we have no way of locating comparable passages of the kind described on the six sets of relevant psychological variables for the forty-five military leaders. Moreover, since social and historical conditions under which a statement is made are likely to color or shape the statement itself, we have no way of establishing the comparability of these conditions for the forty-five leaders across time and space.

Strictly speaking, then, we are left with nonscientific ways of proceeding with our analyses. Although autobiographies, diaries, and memoirs have been relatively rare to come by, we have been able to locate a large collection of biographical studies supplemented by historical works (see the Bibliography). We have subjected our source materials to close scrutiny in search of evidence for the six sets of psychological variables we have identified. The following is a list of the kinds of evidence we sought in connection with each variable.

Nationalism: Feelings of pride in a nation, together with desire and action to improve the status and power of that nation.

Conservative (or Ultra-) Nationalism: Feelings of pride in a nation, together with desire and action to improve the status and power of that nation *at the expense of other nations.* Another word for ultranationalism is imperialism.

Relative Deprivation: Perception of discrepancy between economic values sought and economic values attained.

Love Deprivation: Loss of one or both parents at an early age, the resultant bereavement, and the attempt to compensate for lost love by seeking prominence in military endeavors.

Marginality: Significant or perceptible deviation from commonly accepted norms, whether physical, social, or psychological.

Vanity: Excessive pride in one's own qualities or accomplishments; self-aggrandizement, ostentatiousness, smugness.

Contrary to our expectations, some of the psychological variables lend themselves to relatively unambiguous coding. For example, some form of nationalism characterizes virtually all military leaders; imperialism is a hallmark of Napoleon; relative deprivation stands out in Eisenhower; love deprivation is a driving force of Jackson; marginality is a prominent mark of Nelson; and vanity is manifest in such a figure as Montgomery.

At times the psychological variables were difficult to code, requiring exhaustive searches of the available source materials and judgment calls rooted in extensive prior research.

As explained in Part I, psychology alone does not account for the rise of military leaders; it is a necessary but not a sufficient variable. For military leaders to emerge, it is imperative that psychology coincide with the presence of certain situations. Taken together, psychology and situation propel men toward military careers.

To reiterate, situational variables take four principal forms: (1) family tradition; (2) conditions of open or latent conflict in which elements of power contests are sufficiently salient to be unavoidable; (3) conditions wherein certain personal attributes beyond one's control (for instance, birthplace, urban culture) set the stage for the assumption of leadership roles; and (4) conditions of luck or chance.

Given the difficulties surrounding their analysis, we give our psychological and situational variables more extended treatment than we have accorded other topics in this book, in order to share with the reader the "flavor" of the source materials with which we worked.

The unevenness of the psychological–situational profiles that follow is determined by the unevenness of the data available to us. Our profiles run between two hundred and fifty words and two thousand words, averaging around a thousand words. Nonetheless, as we shall see, they shed light on our substantive concerns.

Finally—we should stress—the profiles that follow are not intended as rounded biographies. Rather, they deal only with the psychological and situational variables discussed above.

Chapter 4

American Leaders

GEORGE WASHINGTON, 1732–1799

George Washington is likely to have experienced love deprivation since (1) his father died in 1743, when George was eleven years old, and (2) for reasons that still remain unexplained, his relations with his mother were always tense and strained. Accordingly, George's older half-brother (by fourteen years), Lawrence, became a surrogate father. As a result, George faced a host of psychological problems: a series of compensatory behaviors modeled after Lawrence.

Although his father and both his older half-brothers had been educated at the Appleby Grammar School in northern England—where one acquired "breadth and polish" (Cunliffe 1958, 29)—George Washington did not proceed beyond an informal elementary schooling. Among the stated reasons for George's nonattendance at Appleby are his father's early death, his mother's possible miserliness, and his mother's possible selfishness in wanting George close to home to help with farm chores.

In any event, Washington's lack of formal education was something to which he and his contemporaries remained sensitive. Thus John Adams was to maintain: "That Washington was not a scholar is certain. That he was too illiterate, unlearned, and unread for his station and reputation is equally past dispute" (quoted in Cunliffe 1958, 31; cf. Flexner 1965–1972, I:23). Moreover:

Nor, of course, does he compare in intellectual preparation and power with such Virginia contemporaries as Thomas Jefferson and James Madison. Years afterward Washington probably felt the lack. He was ill at ease in set debate or abstract discussion. He managed to express himself on paper with a degree of clarity and force, through long practice, and his spelling likewise improved.... We may attribute a little of the constraint on the mature Washington to his awareness of his own intellectual limitations. While still a young man, he was to suffer through his ignorance of the French language, and afterwards he was to refuse an invitation to visit France, on the grounds that he would be embarrassed by having to converse through an interpreter. (Cunliffe 1958, 31–32; cf. Flexner 1965–1972, I:24)

From age six on, George Washington came under the spell of Lawrence, who became a "mentor," a "model," and George's "first and only hero" (see, for example, Freeman 1968, 2–8; Bellamy 1951, 26–27). In addition to his Appleby education, Lawrence was a self-trained surveyor, was appointed adjutant general of the Virginia militia, held a royal commission as a captain in the British regular army, and had been elected to the Virginia House of Burgesses. In every one of these capacities, Lawrence left an indelible imprint on George. Moreover, he introduced George to the aristocratic and influential Fairfax family. And when Lawrence died of tuberculosis in 1752 (and when his wife remarried), George came to inherit Mount Vernon (which Lawrence had inherited from their father). In one way or another, Washington emulated Lawrence throughout most of his adult life.

Washington found surveying congenial because he was mathematically inclined, he enjoyed the company of Lawrence (together, they surveyed the vast Fairfax holdings, as well as other properties), and surveying allowed George to spend time at Mount Vernon and away from Ferry Farm, where his mother presided. By available accounts, Mary Ball Washington was a selfish, possessive, and strong-willed woman; two sources describe her as a "termagant" (Flexner 1965–1972, I:ch. 2; Bellamy 1951, ch. 2). Following Augustine's death in 1743, Mary Ball became increasingly domineering:

For years Mary Ball spared no effort to reduce her son to obedience—an obedience which she tied up with her personal comfort.... Mary Ball did something to him [George] in his childhood which affected his character throughout his entire life. (Bellamy 1951, 18, 19)

Mary Ball Washington did not approve of Martha Custis Washington. "For thirty years after his marriage [in 1759]... she never visited Mount Vernon" (Bellamy 1951, 20).

Later on, Mary Ball became resentful of Washington's achievements and triumphs; she viewed his military and political activities solely as something that prevented him from giving her the attention that she deserved. She complained bitterly and in public; and she died an embittered woman in 1789, at age eighty-three.

Having rebelled against his mother, for his part Washington maintained a stern formality and distance. A biographer writes circumspectly:

The strangest mystery of Washington's life was his lack of affection for his mother. Added years and understanding brought no improvement in his relations with her. As a matter of filial duty he left instructions ... that his mother's calls for money were to be met, but apparently he did not write her even once during the war. He who had so much magnanimity and patience in dealing with human frailty was so much like his mother, in most money matters, that he felt she had been grasping and unreasonable. (Freeman 1968, 514)

As the adjutant general of Virginia, Lawrence had arranged for Washington to join the militia in 1752. Upon Lawrence's death later that year, Virginia was divided into four districts, and Washington was given a district adjutantship. On taking the oath of office, he officially became Major Washington. "To seek to succeed his brother was, for George, a natural ambition" (Freeman 1968, 32).

Washington became a militia officer on the eve of the resumption of hostilities between England and France, culminating in the French and Indian War of 1754–1763. Having taken part in a series of military missions, Washington was promoted to lieutenant colonel in 1753 and to colonel in 1755. But his real ambition lay elsewhere: he wanted a royal commission, with comparable rank, in the British regular army. After all, had Lawrence not been made a captain in 1740, following Britain's declaration of war on Spain?

Between 1753 and 1758 Washington did all he could to make a case for a royal commission with every available British civilian and military official; and he came to equate promotion with "honor":

In other words, he did everything feasible to win preferment ... everything, that is, short of dishonor. There is something unlikable about George Washington of 1753–1758. He seems a trifle raw and strident, too much on his dignity, too ready to complain, too nakedly concerned with promotion. Yet he had real grievances; he was efficient and resolute. His fault lay in saying so too frequently to other people, and in nearly developing a persecution complex as his hopes faded after a promising, almost sensational early start....

With the longing for preferment went the thirst for "honor." Sometimes Washington defined this so as to make it almost synonymous with preferment. All through his adult life Washington was to be closely concerned with his reputation.... Washington needed the solace of public approval. (Cunliffe 1958, 58–59; cf. Flexner 1965–1972, I:121–22, 143–48, 160–62, 171–76, 185–86; Freeman 1968, 138)

His military ambition dashed, Washington resigned from the militia in 1758 and returned to civilian life with "bitterness, resentful of injustice, stung with a sense of unmerited failure" (Flexner 1965–1972, I:223).

Back in Mount Vernon, George Washington compensated for military disappointment by pursuing wealth and social distinction, both symbolized in the acquisition of land. Over the next sixteen years (that is, up until the First Continental Congress), Washington's drive for land was resolute, relentless, almost obsessive.

To be sure, in Colonial America land acquisition was a passion with almost everyone, including Washington's father and grandfather; George himself controlled some six thousand acres by the time he was twenty-one years old. Nonetheless, the 1758–1774 period is different: "In no other direction did Washington demonstrate such acquisitiveness as in his quest for the ownership of land"; a contemporary characterized his pursuit of land as "avaricious" (Flexner 1965–1972, I:289, 303).

By the mid-1760s (partly as a result of his 1759 marriage to Martha), Washington's landholdings had grown to fifteen thousand acres, but he remained unsatiated; he was "operating like a man with twenty hands" (Flexner 1965–1972, I:293). Washington's particular foci of attention were a gubernatorial promise and a royal proclamation authorizing the granting of land to the veterans of the French and Indian War. With considerable resourcefulness, he politicked, maneuvered, petitioned, advertised in newspapers, organized expeditions to survey the lands—all in the hope of expanding his holdings. The coup de grâce, however, came in 1774: "Lord Hillsborough, the Secretary of State [for Virginia], outlawed his claims to thousands of acres by finally ruling that the grants which the Crown had made to veterans of the French and Indian War applied only to regulars" and not to the colonial militia (Flexner 1965–1972, I: 321).

Meanwhile, following Lawrence's example, planter Washington and soldier Washington had developed an interest in politics. Back in 1758, Washington had been elected to the Virginia House of Burgesses, and he had been successively reelected in 1761, 1765, 1768, 1771, and 1774. As the issues of the revolutionary period crystallized—taxation, representation, "the rights of man"—Washington found his sympathies to be with the colonial rebels. In a way, then, the American Revolution intruded on Washington's quest for land.

In 1774, as a burgess, Washington was elected as one of seven Virginia delegates to the First Continental Congress. Having been reelected a delegate to the Second Congress, Washington arrived in Philadelphia wearing his military uniform, the only one in the assembly: "He had brought with him from Mount Vernon a uniform he had worn in the French and Indian War—and now he was wearing it daily, as if to signify to his fellow-delegates that he believed the time had come to take the field"

(Freeman 1968, 217). John Adams wrote Abigail: "Colonel Washington appears in Congress in uniform, and by his great experience and abilities in military matters he is of much service to us" (quoted in Flexner 1965–1972, I:334).

Washington was unanimously named General and unanimously elected Commander-in-Chief. In this capacity, he would command all the regular forces of all the thirteen states—and he would fight those who had denied him commission in the British regular army.

JOHN PAUL JONES (NÉ JOHN PAUL, JR.), 1747–1792

Myth, legend, adventure, and heroism surround the story of John Paul Jones. The facts are more sobering.

John Paul Jones was born John Paul, Jr., in 1747, the son of John Paul, Sr., a gardener. His rise as a naval officer was situational in at least two important ways.

First, raised in Kirkbean on the southernmost point of coastal Scotland, the boy was deeply influenced by the sights and sounds of the sea and sea vessels. Routinely listening to "the stories of the mariners" and his "sailor-teachers," he developed an "early talent for seamanship" (De Koven 1913, I:6). As a youngster, "often he would launch his little boat upon the waters, calling out to his imaginary crew in words of authority and command—the hero in miniature, a prophetic and engaging picture!" (De Koven 1913, I:7). In short, John Paul developed an early love of the sea: "The sea! That was what young Paul wanted from the first, as his road to distinction" (Morison 1959, 10).

Second, John Paul's life coincided with "the day of England's imperial expansion, and many of her sons were moved to leave their home shores on errands of glory or gain" (De Koven 1913, I:7). William, John Paul's eldest brother, early departed for America, as did other family members and friends. The lure of America was irresistible; it became John Paul's enchanted land.

Ambitious from an early age, John Paul was determined "very young" that he "did not intend to be a gardener, farmer, or fisherman" (Morison 1959, 9). Having been born "in obscurity and poverty" (Morison 1959, 3), having experienced relative deprivation in childhood, he was bent on overcoming it.

Apprenticed to a local merchant shipper, John Paul sailed at age twelve (in 1759) as a cabin boy on a ship to Virginia, where he visited his brother William, a tailor in Fredericksburg. The apprenticeship ended in 1766, and for a time John Paul entered the slave trade, after which he headed back for Scotland. In April 1770, in Tobago, his "ungovernable temper" (Morison 1959, 19) led him to flog the ship's carpenter (one

Mungo Maxwell, a Tobagan) for laziness, incompetence, and insubordination, leading to the latter's death a few months later.

When John Paul reached Scotland in November 1770, he was arrested on murder charges but was released on bail. In September 1772, he purchased a vessel and made several trips to Tobago in search of evidence (in the form of affidavits from British sources who had known the carpenter) to clear his name. During his final trip, in December 1773, he killed the leader of a mutinous crew in self-defense (or so he claimed). He fled the island and changed his name to John Paul Jones.

Meanwhile, in America, Jones had met some southern leaders of the revolutionary movement—most notably the shipbuilder Joseph Hewes—who persuaded him to go to Philadelphia. There in the fall of 1775 he received a commission in the Continental Navy. Over the next four years, he scored a series of daring naval maneuvers that made his name a household word and established him as the "Father of the American Navy."

In 1779–1780, Jones headed a French naval expedition against Britain with great success. Between 1781 and 1787, he traveled back and forth between America and France in search of glory and adventure. In 1788, on the advice of Thomas Jefferson, then ambassador to France, Jones accepted the offer of Catherine the Great to join the Russian navy as a rear admiral. While in Russia, he was accused of sexual assault upon a young woman. In 1790, he returned to France a frustrated and broken man, where he died two years later.

Other than being a fugitive from justice and a potential criminal, Jones was primarily an adventurer in a ceaseless search for personal glory. His loyalty was to himself alone—not to Britain, America, Russia, or France. In addition to being ambitious, he was vain, arrogant, contentious, sarcastic, and contemptuous (Mackenzie 1846, II:Ch. 20). He was "a colossal egotist" and an unceasing seeker of glory (Morison 1959, 92 et passim; cf. Lorenz 1943, passim). According to another biographer, Jones "loved glory as [Horatio] Nelson loved it: openly, unblushingly, seeking it at the cannon's mouth.... His soul, ardent as strong, was inflamed by an unlimited thirst for glory" (De Koven 1913, II:423, 424). "He pretended total indifference to fame, but he took every possible means to place a far from modest estimate of himself before the public of two continents" (Morison 1959, 3–4).

Against these assessments, we should note that at times Jones considered himself a Citizen of the World, claiming to fight for liberty wherever it was endangered (Morison 1959, 32, 88).

ANDREW JACKSON, 1767–1845

Andrew Jackson's early experiences created a sense of both relative deprivation and love deprivation, setting him on a course of seeking vindication throughout his life. A biographer puts it this way:

Andrew Jackson was more than a symbol; he was a vital force. As the force was rarely at rest, so the man was rarely at peace. Beneath the aggressiveness, the boldness, the quick temper lay deep uncertainties rooted in his precarious back country upbringing and in events that left him an orphan at the age of fourteen. Throughout his life he felt a need to prove himself, to triumph over enemies he believed were assaulting his reputation. Even in victory, he never felt victorious. The more his reputation grew, the more he feared that some conspiracy, some cabal was working to diminish his standing with the people.

Confronted by death at an early age, Andrew Jackson spent his life trying to prove his right to survival. This quest profoundly influenced his own destiny and that of the nation. Jackson's personal correspondence reveals an intense inner turmoil. I have tried to understand the nature and origins of this emotional turbulence, for it affected all his major decisions whether as frontier general, treaty negotiator, party leader, or chief executive. To overlook these feelings is to miss the essence of the man and the basis for his popular appeal. (Curtis 1976, ix-x)

Jackson grew up in a rough and chaotic environment. Raised in Ulster, Northern Ireland, his parents emigrated to the New World in 1765. An itinerant worker, Jackson's father died in a work-related accident in 1767, when his mother was seven months pregnant with Andrew.

Life on the frontier was laden with conflict and violence; Jackson became "wild" and "mischievous"; his "behavior reached extremes"; he never properly learned the English language; he developed no intellectual interests (Curtis 1976, 7).

Following his father's death, Jackson's mother went to live with her sister, Jane Crawford. The Crawfords had eight children; and Andrew had two older brothers. So, for twelve years, Jackson experienced both the support and the turmoil that a large household engenders.

Nevertheless, Jackson was essentially a bully; he had a bad temper; he terrorized his cohorts; he was reckless; he was extremely sensitive and defensive (especially about his youthful habit of slobbering); he was filled with rage and anger (Curtis 1976, 7–8; Remini 1977, 6–10).

Jackson joined the Continental Army at age thirteen. His older brothers, Hugh and Robert, died in the Revolutionary War in 1779 and 1781, respectively. Jackson's mother also died unexpectedly in 1781. At age fourteen, Jackson found himself rootless and utterly alone. He tried to board with relatives but they were alienated by his rough and aggressive behavior. "Homeless and friendless" were words Jackson used to describe this period of his life (quoted in Curtis 1976, 12). In 1785, much to everyone's relief, Jackson left for Salisbury, North Carolina, to study law. In Salisbury, too, he was "the most rolling, rollicking, gamecocking" person around (Curtis 1976, 13).

From 1788 to 1796, Jackson practiced law in Nashville, Tennessee, but social position and status eluded him:

He lacked the family connections, property, and political standing to enter the lists of the tidewater aristocracy. Yet aristocrat he wanted to be.... In the unsettled society of the new colony, Jackson expected his legal skills to compensate for the lack of family connections and education. (Curtis 1976, 21)

Even though he held a very low opinion of politics and politicians, Jackson turned to politics as a means of enhancing his social status, first as a U.S. representative, then as a U.S. senator, and eventually as U.S. president. Through it all he was moved by the relative deprivation and the love deprivation that had marked his early life, and by a continuing need for vindication.

ROBERT E. LEE, 1807–1870

By all accounts, Robert E. Lee was a man of modesty, simplicity, humility, and self-denial (see, for example, Jones 1874, passim; Freeman 1934–1935, passim). Nonetheless, his rendezvous with the military was virtually predetermined.

First, Lee was born to a military family. His father, Henry Lee (1756–1818), was a military man who had fought in the Revolutionary War, rising to the rank of general in 1786. He regularly regaled the family with sagas of military adventures.

Second, Lee's family home of Alexandria was located near Mount Vernon. This meant, inevitably, glorification of the Revolutionary War, George Washington, and military service. In fact, "amid these surroundings, Washington was part of the life of Robert Lee from earliest childhood.... In the home where Robert was trained, God came first and then Washington" (Freeman 1934–1935, I:22).

Third, in the summer of 1814, when Lee was seven years old, the British captured Alexandria, an event that made a deep impression on the young boy. He personally heard the sounds of war; he saw Redcoats everywhere. "A grim early memory it was for a soldier's son, destined to be a soldier himself!" (Freeman 1934–1935, I:30).

Beginning in the late 1780s, Henry Lee developed serious health problems, compounded by financial difficulties. His second marriage to Ann Hill Carter—Robert's mother-to-be—in June 1793 considerably improved his spirits, but his troubles were not over. In the spring of 1809, when Robert was two years old, Henry Lee's real estate was confiscated for a bad debt, and he was imprisoned for nearly a year—events that made an indelible impression on the young child (see Freeman 1934–1935, I: ch. 1).

Given Henry Lee's circumstances, all responsibility fell on the mother:

Over it [the family home] presided his mother, charged for the rest of her days with the entire care of her five children, their finances, their religious training, and their education. Physically, it overtaxed her, but spiritually she was equal to it. (Freeman 1934–1935, I:22)

The mother inculcated the children with the axioms of frugality, responsibility, self-discipline, self-control, and self-denial. Robert Lee would say later in life that he "owed everything" to his mother (quoted in Freeman 1934–1935, I:23).

Unhappily, Henry Lee died in 1818 and the mother's health began to deteriorate as well. Beginning at the tender age of eleven, Robert spent a great deal of time caring for his sick mother.

In 1825, Robert E. Lee applied—and was admitted to—West Point; he graduated second in his class in 1829. Learning of the worsening condition of his mother, he rushed home to resume "his old duties as a nurse" (Freeman 1934–1935, I:87). As fate would have it, she died on July 10 of that year, freeing Robert E. Lee to pursue his military career with singular determination.

Given Lee's life experiences, it is not surprising that a biographer considers "duty" as the keynote of his life and career:

If asked to name in a single word the controlling principle of General Lee's life, we should unhesitatingly answer, DUTY. Whether as a youth meeting his obligations to his aged mother, and passing through the Military Academy without a single demerit; or serving in the United States Army; or directing the forces of his native South; or quietly working in the college at Lexington for the good of the young men of the country—*duty* was the star which guided him throughout his eventful career.... "Duty [Lee said] is the sublimest word in the English language." (Jones 1874, 133; capitalization and italics in original)

WILLIAM TECUMSEH SHERMAN, 1820–1891

Born in Lancaster, Ohio, a military career was virtually thrust upon William Tecumseh Sherman. (He was named Tecumseh, by the way, because his father admired the Shawnee Indian chief of that name, who had played an important role as a noted warrior in Ohio frontier history.)

William's father, Charles Robert Sherman, and Thomas Ewing, both attorneys, became not only neighbors but intimate friends as well through their many travels together. "They became as intimate as brothers, sharing beds, meals, and lawbooks" (Merrill 1971, 16). Ewing was a wealthy man who subsequently went on to become an influential United States senator.

Charles Robert died unexpectedly in 1829 (when William was only nine years old), leaving his wife and eleven children bankrupt and penniless. The children were farmed out to family and friends, leaving William to feel that his mother had abandoned him as well (Marszalek 1993, 10). Relative deprivation and love deprivation left lasting impressions on Sherman's life.

As chance would have it, William's lot fell with Senator Ewing, who informally adopted Sherman and treated him as a member of his own family. (He forewent legal adoption in order to allow Sherman to retain his birth name.)

In 1834, Ewing instructed Sherman (then fourteen years old) to prepare for West Point. In 1836, through Ewing's connections, Sherman entered West Point (which he did not particularly enjoy), and graduated in 1840. This experience created in Sherman ambivalent feelings toward his foster father. However, Sherman fully embraced the profoundly conservative philosophy that Ewing expounded.

Marszalek (1993) has argued that the chaos of Sherman's childhood generated in him a dread of social instability and a "passion for order" (the book's subtitle). A rootless childhood, shaped by his father's early death and subsequent separation from his mother, prompted a series of struggles for order. Efficient military pursuits became the sole driving force of Sherman's life. He preferred the company of his men to that of his wife and children (cf. Liddell-Hart 1978, 425-31 et passim).

Marszalek continues:

[Sherman] entered adulthood with many unresolved childhood conflicts. He felt rootless, the result of his ambiguous feelings toward his natural and his foster families. Only the army seemed to provide the sense of belonging he yearned for, but West Point life had been frustrating. The new second lieutenant emerged into his new life with old relationships unsettled and destined to influence his future more than he ever realized at the time. (Marszalek 1993, 29)

As a result, throughout his military career, Sherman experienced periods of depression and restlessness (Liddell-Hart 1978, passim; Marszalek 1993, passim). He compensated for his inadequacies by developing a colossal self-concept; he saw himself "as a character on the world's stage" (Liddell-Hart 1978, 200). Another biographer notes:

There runs through General Sherman's correspondence, and very noticeably in his *Personal Memoirs*, an almost overbearing egotism and an overdeveloped sense of dignity, easily offended. Such emphasis on self, with attendant braggadocio, usually indicates pronounced egocentrism, and in many cases arises from a sense of inadequacy. It also causes the subject to indulge in self-pity and to keep ever fresh in his mind all the past failures and grievances. With self-pity comes great mental anguish, periods of gloom and despair, a tendency to exaggerate the difficulties of problems and to resort to rationalizations in moments of strain and crisis. (Walters 1973, 16)

Sherman was a savage, ruthless, and merciless warrior—so much so that one author (Walters 1973) titled his biography of Sherman *Merchant of Terror*. Sherman became so mercurial, volatile, and temperamental that he was commonly called "Crazy Sherman"—indeed, at least one biographer believes that there was an element of actual insanity about Sherman (Merrill 1971, 176, 178, 184–88).

ALFRED THAYER MAHAN, 1840–1914

Alfred Thayer Mahan is one of the most controversial figures in the annals of United States naval history. A biographer encapsulates his subject in the following terms:

It is the portrait—warts and all—of a historian, strategist, tactician, philosopher, Episcopalian, theologian, diplomat, imperialist, mercantilist, capitalist, Anglophile, patriot, Republican, racist, Social Darwinist, journalist, polemicist, naval reformer, adviser to presidents and legislators, teacher, academic administrator, social climber, egoist, introvert, swain, husband, and father. (Seager 1977, xi)

Mahan was born and raised on the grounds of the U.S. Military Academy at West Point, where his father, Dennis Hart Mahan, was dean of the faculty and professor of military engineering. The family motto was the same as West Point's: Duty, Honor, Country. At a very early age, Mahan internalized an appreciation of the military profession. A military career was preordained for him.

A voracious reader of military literature since childhood, Mahan became particularly fascinated with the career of Horatio Nelson, an admiring biography of whom he eventually published (Mahan 1897). Through Nelson, Mahan became convinced of the influence of seapower in military matters, producing his most famous book, *The Influence of Seapower Upon History, 1660–1783* (Mahan 1890). Accordingly, Mahan entered the U.S. Naval Academy in 1854, graduating four years later.

Mahan's personal traits include vanity, egotism, racism, and imperialism—all justified under the guise of Christianity (he was a devout Episcopalian). A biographer writes:

At the heart of Mahan's problems . . . was his ill-concealed vanity. The fact of the matter is that he considered his appearance, his mentality, his morality, and all of his own works, ideas, and attitudes to be vastly superior to those of the common run of mankind.... He could neither understand nor tolerate anyone who disagreed with him. Such people were either perverse or malevolent. (Seager 1977, 27; cf. Puleston 1939, 25)

Although born in New York, Mahan's family came from Virginia, and he typically looked upon African-Americans as "niggers" and "darkies" (Seager 1977, 29). According to Mahan, the "English-speaking family" was united by "race patriotism" (quoted in Seager 1977, 350).

An exponent of Social Darwinism, a believer in White Man's Burden, a champion of the Monroe Doctrine, and a staunch supporter of the Spanish–American War, Mahan felt called upon to spread Anglo-Saxonism throughout the world (Seager 1977, passim; Puleston 1939, passim).

Mahan's beliefs and actions were informed by a literal reading of the New Testament. A biographer writes:

Religion was always an important force in the moulding and expression of his personality and character. It was a major influence in the later formation of his military and political ideas as well as his philosophy of history. (Seager 1977, 6; cf. Puleston 1939, ch. 1)

Finally, we should note, from age sixteen on, Mahan experienced periods of mental depression (Seager 1977, 12 et passim); and from age twenty on, he developed a drinking problem (Seager 1977, 33 et passim).

JOHN J. PERSHING, 1860–1948

John J. Pershing's encounter with the military was fortuitous in several ways.

Born a year before the beginning of the Civil War, Pershing was brought up in the atmosphere of great racial strife and its aftermath.

In the panic of 1873, the family's mercantile business and landholdings were wiped out. Life became very harsh for the family of eight. Pershing experienced relative deprivation and resolved to improve his socioeconomic status.

Pershing's early ambition was to become a teacher or a lawyer. Having gone through Kirksville Normal School near his hometown of Laclede, Missouri, he obtained a teaching certificate. The only opening he could find was for a teacher in Laclede Negro School, where he confronted prejudice from both blacks and whites. The black parents boycotted the school, demanding a black teacher. The white parents shouted outside his classroom: "Nigger! Nigger!" (Vandiver 1977, I:13–14).

The whole course of Pershing's life changed in July 1881, when quite by accident he came across a newspaper announcement of a competitive examination for the appointment of a cadet at West Point. His parents objected, but his younger sister Elizabeth helped him cram for two weeks, passing the examination later that month. "As John rationalized it, ... West Point would provide a first-rate education at government expense, and he was privileged to resign from the military life after graduation" (O'Connor 1961, 3). As an avenue of social mobility, Pershing attended West Point from 1882 to 1886. Feeling out of place at first, he gradually adjusted to military routine.

Still wavering in the choice of a career, while serving as a military instructor at the University of Nebraska, Pershing managed to get a bachelor of laws degree in 1893. He was admitted to the Nebraska bar later that year.

In 1897 Pershing joined the 10th Cavalry—which was made up largely of blacks—and confronted the racial issue all over again:

Unpopularity won a nickname for Pershing, one born in racist contempt. The cadets knew he belonged to the 10th Cavalry and so began to call him "Nigger Jack." In time it softened to "Black Jack," but the intent remained hostile. (Vandiver 1977, I:171; cf. Smythe 1973, 44)

By now a man of towering ambition—and possessed of an insatiable desire to make a name for himself—Pershing resolved to transcend his life circumstances. The Spanish–American War and World War I provided the opportunities. Military success, in turn, nurtured a highly egocentric personality.

GEORGE C. MARSHALL, 1880–1951

George C. Marshall was born in a time of socioeconomic turbulence to a family with a military background. He experienced relative deprivation and physical marginality in childhood. As a result, he developed a strong urge to excel.

George's father, George Sr., was a Civil War veteran who shared with the family many anecdotes and stories of the great conflict. George's older brother Stuart was a graduate of Virginia Military Institute, an institution of considerable regional repute. As a youngster, George daydreamed about becoming a soldier:

The boy was hardly in his teens when he determined to be a soldier. His sister Marie could never remember a time when he was not dreaming of entering the Virginia Military Institute. He was ten when his brother entered the Institute, and there were Stuart Bradford's letters home to remind him of the glories of the place, and there were the family traditions and all of Virginia history to tempt him toward the college at Lexington. He imagined battles when he was young, himself a commander of invisible armies in the garden behind his house, where he threw up a military tent, made from a blanket thrown over the branches of an apple tree. (Payne 1951, 14)

Marshall's early life coincided with the heightening of urbanization, industrialization, strikes, and violence—and the concomitant need for continuity and order. More immediate was the family's loss of financial security in the early 1890s. George Sr., a merchant, had invested a small fortune (from the sale of coal and wood) in land and a hotel in a Virginia resort. Soon thereafter, the landboom in the area collapsed and the hotel burned down. Almost overnight, George Sr. lost virtually everything he had worked for. Marshall would later refer to this decision by his father as "the great mistake of his life"—one that forced the family "to economize bitterly" (quoted in Stoler 1989, 4–5). The mother's landholdings became the only source of family income. As a result, Marshall developed "a passion for solvency which he never lost" (Pogue 1963, 35).

Marshall experienced physical marginality in childhood:

He remembered being particularly awkward at the age [ca. seven] he went first to public school. He was rather tall and slender, snub-nosed, with a mop of sandy hair parted, as the fashion was, squarely in the middle. He remembered having big feet, though in fact his feet were not large. He was quiet, shy, and perhaps unusually serious. He remembered that people "made fun of me a great deal." Though he knew everybody and ran happily with the gang, he was apparently even then learning the reserve for which he was afterwards noted. (Pogue 1963, 20)

Marshall's relationship with his mother was warm and affectionate, while his father and siblings remained distant and aloof:

Perhaps most notable ... was the close relationship Marshall developed with his mother, a relationship encouraged by the society's mores, as well as by his status as the youngest child. She was a quiet, affectionate, and supportive woman, and her younger son seemed to hold a special place in her heart. When Stuart and Marie went off to school, only he remained at home, and he was thus to an extent raised as an only child. He later admitted that his mother had spoiled him. More important, she became his "confidante in practically all my boyish escapades and difficulties." Clearly she was the most important person in his childhood and by his own recollection a "constant and lasting influence on my life."

His relationships with his father and siblings were not as close or warm. To his older brother and sister, he was primarily a younger embarrassment and annoyance, and he was not intimate with either of them. Nor was he close to his father, a man who ... was cold and aloof at home, as well as highly critical of his younger son. ... Marshall's recollection in later years ... was that his father simply preferred older brother Stuart to him. (Stoler 1989, 5; cf. Pogue 1963, 27–30)

Marshall's possible feelings of rejection and low self-esteem, together with his poor performance at school, set in motion a series of compensatory mechanisms:

Ashamed and humiliated by his academic performance and the treatment he received from his peers, distant from his father and his siblings, young Marshall responded by seeking approval and companionship elsewhere. . . . Marshall also responded by developing some protective and distinctive character traits that complemented parental values and would remain key components of his adult personality. Most notable in this regard were his shyness and reserve, his desire to excel as a means of proving his worth to his critics, and his insistence on avoiding the appearance of failure or flaw, an insistence that later became an important aspect of his leadership ability. (Stoler 1989, 5–6)

Given these circumstances, given recent memories of the Civil War, and given the romance associated with soldiering, by his mid-teens Marshall had resolved to join the military. In 1897, over the objections of his parents, he entered the Virginia Military Institute, and graduated four years later. In 1901 he volunteered for service in the Philippines.

To recapitulate, Marshall's personal, economic, social, and psychological experiences generated an intense ambition to succeed in the face of difficult odds (see Pogue 1963, passim; Payne 1951, passim; Stoler 1989, passim). To cap it off, Marshall is the only military leader in our sample to have won a Nobel Peace Prize for the plan that bears his name.

DOUGLAS MacARTHUR, 1880–1964

One of Douglas MacArthur's biographers has captured some outstanding traits of his subject in the following fashion:

He was a great thundering paradox of a man, noble and ignoble, inspiring and outrageous, arrogant and shy, the best of men and the worst of men, the most protean, most ridiculous, and most sublime. No more baffling, exasperating soldier ever wore a uniform. Flamboyant, imperious, and apocalyptic, he carried the plumage of a flamingo, could not acknowledge errors, and tried to cover up all his mistakes with sly, childish tricks. Yet he was also endowed with great personal charm, a will of iron, and a soaring intellect. Unquestionably he was the most gifted man-at-arms this nation has produced. He was also extraordinarily brave. His twenty-two medals—thirteen of them for heroism—probably exceeded those of any other figure in American history. (Manchester 1978, 3; cf. Gunther 1951, 12, 23–24)

MacArthur was born into a military family and entirely immersed in military life from the earliest childhood. A military career was charted for him by his father.

MacArthur was born on an army post near Little Rock, Arkansas, and he lived in various military posts most of his youth. It is "a matter of family record that, as a child, he heard the drum beats of hostile Indians. Camp life, army life, dominated all his early years, and it never occurred to him to be anything but a soldier when he grew up" (Gunther 1951, 32; cf. Hunt 1954, 4). MacArthur himself noted that it was at a military outpost as early as age five that "I learned to ride and shoot even before I could read or write—indeed, almost before I could walk and talk" (MacArthur 1964, 15). A manifest element of exaggeration notwithstanding, the statement reflects MacArthur's early fascination with military life.

Douglas's father, Arthur MacArthur, Jr. (1845–1912), was commissioned a first lieutenant in the Union Army in 1862. In 1863, he was awarded the Congressional Medal of Honor for bravery in the Civil War. Having been promoted to brigadier general, he fought in the Spanish–American War. In 1900, he was named military governor of the Philippines. In all of these capacities, MacArthur successfully emulated his father. The "deeds and glories" of the father "filled the imagination and the memory of the boy Douglas" (Hunt 1954, 6).

In his youth MacArthur "idolized his father" (Manchester 1978, 4). As early as 1893, when he was thirteen, he overheard his father remark to his mother: "I think there is the material of a soldier in that boy" (quoted in Manchester 1978, 44).

"As far back as he could remember, his father had expounded to him the glories of West Point and had gone about the task of lining up an appointment" (Hunt 1954, 15). Accordingly, as preparation for West Point, MacArthur entered the West Texas Military Academy in 1893, and graduated four years later.

After graduating as valedictorian from the Texas academy, MacArthur tried to secure an appointment to West Point. Having been denied presidential appointments from both Grover Cleveland and William McKinley, despite his impressive dossier, endorsements from thirteen assorted governors, senators, and congressmen, MacArthur finally secured an appointment through Congressman Theabald Otjen. However, he flunked the physical examination. One year later, and after working with a physician to correct the curvature in his spine, MacArthur entered West Point in 1899 at age nineteen, graduating in 1903. (On this paragraph, see Manchester 1978, 44–47; James 1972–1985, I:62–63.)

Following in the footsteps of the father, in 1936 MacArthur became Philippines field marshal. In 1942, he was awarded the Congressional Medal of Honor. In 1944, he was promoted to five-star general. And in the postwar period, he became the virtual ruler-dictator of Japan—indeed, he considered himself "a sovereign" (quoted in Manchester 1978, 471; cf. Gunther 1951, 97).

MacArthur's legendary vanity and egotism trace back to childhood:

The father's career and character put a mark on MacArthur that nothing has ever effaced, but he was the child of his mother too. He adored her; she adored him. She deliberately brought him up to think that he was a child of destiny and her influence was illimitable; much of his ambition comes from her, his personal force, and his faith in the future. (Gunther 1951, 32–33)

As MacArthur's career progressed, vanity, narcissism, arrogance, conceit, and megalomania came to constitute a way of life for him. Gunther (1951, ch. 1) called him the "Caesar of the Pacific" and noted MacArthur's "obsession that he is utterly indispensable" and his conviction that he has a " 'divine' mission" (Gunther 1951, 5, 6).

Similarly, Manchester (1978) titled his biography of MacArthur *American Caesar*, noting that he was "conceited and ostentatious," that he "yearned for public adulation," that he "demanded that he be revered" (Manchester 1978, 3, 7). Manchester quoted Clare Boothe Luce's observation that MacArthur "plainly relished idolatry" (Manchester 1978, 6). Manchester concluded: "Most of all, however, MacArthur was like Julius Caesar: bold, aloof, austere, egotistical, willful" (Manchester 1978, 8).

MacArthur's vanity was reflected even in his approach to religion:

His belief in an Episcopalian, merciful God was genuine, yet he seemed to worship only at the altar of himself. He never went to church, but he read the Bible every day and he regarded himself as one of the world's two greatest defenders of Christendom. (The other was the Pope.) (Manchester 1978, 3; cf. Gunther 1951, 75)

WILLIAM F. HALSEY, JR., 1882–1959

Halsey's service in the United States Navy was situational and virtually predetermined by his family circumstances.

"Halsey's fighting spirit and love of the sea had ample family precedent" (Potter 1985, 19). By his own account, Halsey's remote ancestors had been "seafarers and adventurers" (Halsey and Bryan 1947, 2; cf. Merrill 1976, 6–7).

More immediately, Halsey's father, William F., Sr., was a Naval Academy graduate and a career naval officer who retired with the rank of captain. As a youngster, Halsey was brought up on the grounds of Annapolis, where his father was an instructor. Accordingly, it was only natural that Halsey would want to follow in the steps of his father (see Halsey and Bryan 1947, 2–3; Merrill 1976, 6; Potter 1985, 19–20).

It was probably for the same reason that Halsey was not particularly ambitious to advance himself in higher office. He seems to have accepted that he would spend his life in a naval career, and promotions would come in stride (cf. Halsey and Bryan 1947, passim). In fact, were it not for the two world wars, he would have probably remained a middle-rank officer.

Accordingly, Halsey's daring military exploits (for which he earned the nickname "Bull Halsey") must have come as a series of surprises. Similarly, it was unexpected that, for his service in defending New Zealand and Australia, in 1944 King George VI would appoint him an Honorary Knight Commander of the Order of the British Empire (Halsey and Bryan 1947, 192; Merrill 1976, 119).

Finally, we should note an element of mysticism about Halsey. He was highly superstitious in general, and about the number thirteen in particular. At one point, he wrote, "we were appalled to find that not only had we been designated Task Force 13, but our sortie had been set for February 13, a Friday" (Halsey and Bryan 1947, 97). The word "jinx" appears with some regularity in Halsey's language (Halsey and Bryan 1947, 85–88 et passim).

GEORGE S. PATTON, 1885–1945

George S. Patton was the descendent of a California/Virginia family with a long tradition of military service. His choice of a military career was entirely situational.

Even as a boy George Patton had wanted to be a general and lead troops in battle. His prepossession with the military is understandable because his Virginia grandfather and seven great-uncles had been Confederate officers. His grandfather, a V.M.I. [Virginia Military Institute] graduate, was killed at Cedar Creek. . . . He was raised in an atmosphere of adulation for those in the family who had fallen in the War Between the States, with strong overtones of pride in their Virginia ancestry.

His family said that, as a young boy, he used to go around the house with a wooden sword saying, "Lieutenant General George S. Patton, Junior." Never, from his earliest childhood, did he wish to be anything else. All that he did throughout his life was aimed at becoming a combat general. (Semmes 1955, 3–4)

Another biographer notes:

[He] determined . . . at a tender age [ca. eight] to become a soldier, the profession for which he was literally destined by tradition and genetics—ordained, so to speak, by the chromosomes of his ancestors. He was the scion of southern gentry and the frontier aristocracy of California, the decadence and sophistication of his Virginian heritage balanced by his forebears from Southern California, a simple and commoner but hardier and more virile stock. (Farago 1964, 44)

It was only natural—and consistent with the family tradition—that Patton would attend Virginia Military Institute (1903–1904) in preparation for West Point (1904–1909). On January 18, 1909, shortly before graduation, he wrote Frederick Ayer, his future father-in-law:

It is hard to answer intelligibly the question: "Why I want to be a soldier." For my own satisfaction I have tried to give myself reasons but have never found any logical ones. I only feel it inside. It is as natural for me to be a soldier as it is to breathe and would be as hard to give up all thought of it as it would be to stop breathing. (quoted in Semmes 1955, 21–22)

Patton was a student of the Civil War—particularly its great leaders—an interest that lasted throughout his lifetime. His studies taught him the maxim that "War will be won with Blood and Guts alone" (quoted in Semmes 1955, 3).

Patton was easily one of the most colorful military leaders we have encountered. He was vain, flamboyant, fiery, flashy, outspoken, profane, capricious, moody, and temperamental. A biographer writes: "He was first and always an exhibitionist. He played to the galleries. His dress and pearl-handled guns were largely stage props that became his trademark in the public mind" (Semmes 1955, 15). Another biographer notes: "Patton was always interested in glory, adulation, recognition, and approval. He believed passionately in the virtue of becoming well and widely known. What he wanted, above all, was applause" (Blumenson 1972–1974, I:8). Patton not only admired Napoleon, he considered himself Napoleon incarnate (Blumenson 1972–1974, passim; Essame 1974, passim; Farago 1964, passim; Semmes 1955, passim).

Finally, we must note an element of mysticism about Patton. A devout Episcopalian, "His thoughts, as demonstrated daily to those close to him, repeatedly indicated that his life was dominated by a feeling of dependence on God. . . . He turned to God for comfort in adversity and . . . [gave] thanks in success" (Semmes 1955, 6). Another source notes: "There was Patton the mystic, who believed that he had been a soldier in previous lives, and that he would be born again—as a soldier. . . . He felt a kinship with the great soldiers of antiquity that was strong enough to persuade him that he had actually lived before" (*Army Times* Editors 1967, 11–12, 22).

CHESTER W. NIMITZ, 1885–1966

Nimitz's encounter with the U.S. navy was entirely situational. First, however, a word about his birth and childhood.

Chester Nimitz, Sr., married Anna Henke in March 1884. He died unexpectedly in August 1884 and Chester W. Nimitz was born on February 24, 1885. In 1890 Anna married William Nimitz, Chester Sr.'s younger brother, who simultaneously became an uncle and a stepfather. Young Chester grew close to his stepfather, who treated him with great care and affection. He was even closer to his "wonderful white-bearded [paternal] grandfather," as he described him (quoted in Potter 1976, 26). As a result, he hardly noticed the loss of his father. Moreover, the grandfather regularly regaled him with endless stories about his experiences in the German merchant marine.

Nimitz's maternal relatives were in the meat business; his stepfather and grandfather were hotel keepers. Nimitz inclined toward neither. At age fifteen he was contemplating his future when a pivotal incident occurred:

In the summer of 1900, Chester for the first time caught a glimpse of an opportunity to broaden his experiences. From Fort Sam Houston, outside San Antonio, came Battery K Third Field Artillery, for training and gunnery practice in the hiss across the Guadalupe from Kerrville [where Chester went to school]. On their way to join Battery K, Second Lieutenants William M. Cruikshank and William I. Westervelt, both brand new West Point Graduates, stopped at the St. Charles Hotel [a family establishment]. Chester was much impressed by their military bearing, their fitting new uniforms, and above all their air of worldly sophistication.

To Chester of course army officers were no novelty. From earliest infancy he had been bounced on their knees at his grandfather's hotel. What fascinated him about Lieutenants Cruikshank and Westervelt was that, with all their fine bearing and polish and their impending responsibilities in the Army, they were only a little older than himself. They had been plucked but recently from humdrum situations like his own and had been educated and launched on a career of travel and high adventure, all at no cost to their families, all for the asking—plus, of course, hard study and stiff competition.

Young Nimitz had no fear of hard study or competition. For him, at last, a door seemed to be opening. All afire with hope and anticipation, he applied to Congressman James Slaydon to take the West Point examination. Slaydon shut the door. He said that all his appointments for the Military Academy were filled. ... Then Congressman Slaydon reopened the door a little. "But," he said, "I have an opening for the U.S. Naval Academy. Are you interested?"

Chester had never so much as heard of the Naval Academy, but he swallowed his disappointment and said yes, determined to seize any opportunity to get an education. (Potter 1976, 29–30)

Nimitz passed the entrance examination in April 1901 and reported to the Naval Academy in September of the same year.

CLAIRE LEE CHENNAULT, 1893–1958

Claire Lee Chennault's early life was dominated by two sets of circumstances: (1) attempts to please his driven father, and (2) attempts to compensate for the loss of his mother and stepmother. In addition, he appears to have had a dual personality.

Chennault's father, John Stonewall Jackson Chennault, was a landlord and a cotton grower given to a rugged physical life style. He prescribed for Claire love of outdoors, running, jumping, hunting, fishing, and horseback riding. This meant for Claire that:

[D]uring those childhood years he began to show the inner tension that drove him throughout his life. More than the others, he seemed driven to win, compelled to outdo. In part he was trying desperately to please a hard father, a man who could bend a sixpenny nail into a staple with his bare hands and was known to believe that "second place is not worth a damn." ... No doubt he was also simply testing himself, seeking to find the limits of his mind and body and personality. (Byrd 1987, 10)

Chennault's mother, Jessie Lee, died in 1901, setting in motion feelings of love deprivation:

Claire was then eight, and even before that time his mother's pregnancies and illness had caused him to live for long periods with either Dr. and Mrs. [William] Lee [maternal grandparents] or his mother's sister Louise Lee Chase. Aunt Lou, a strong woman who ruled her household with a firm hand and the sure authority of the Bible behind her, now found room in her heart and home for the growing Claire, tucked him in among her own sons, and set before him the example of her own fighting spirit and indomitable will. (Byrd 1987, 7–9)

Two years later, Chennault's father "married Lottie Barnes, who had been Claire's teacher—and a much loved one—at the local school" (Byrd 1987, 9). Developing a very strong attachment to his stepmother, Chennault's feelings of love deprivation were considerably assuaged—but only temporarily. The next blow came when he was sixteen years old:

On 28 November 1909, just as Claire was completing his second semester [at Louisiana State University], Lottie Barnes Chennault died. "I was alone again," he wrote in his memoirs, "and really never found another companion whom I could so completely admire, respect, and love." ... [H]er death, coupled with that of his own mother, left a void in his life that was never totally filled. (Byrd 1987, 14; cf. Chennault 1949, 4)

The event was so traumatic that Chennault decided not to return to LSU. Instead, he set out in search of a career that would help fill the void. At first he contemplated West Point or Annapolis, but he had neither a strong academic record nor the necessary political connections. Instead, he obtained a teaching certificate from the State Normal School at Natchitoches and taught in Louisiana for a time. In 1917, he joined the United States Army and entered the Army's Officer Training School. Not finding the army to his liking, he joined the air force and led the Flying Tigers. Given these circumstances, Chennault's claim in his memoirs (1949, 3) that "All my life I wanted to be a soldier" must be qualified.

Biographer Martha Byrd (1987) portrays Chennault as a man with a dual personality. She imputes to him "a character and personality so full of contradictions that at times there seem to have been two men competing within the same life for the ascendancy of two different value systems" (Byrd 1987, ix). Accordingly, having a Chinese wife, Chennault was caught between the cultures of the East and the West. He was alternately warlike and peaceful, vain and humble, arrogant and unassuming, abrasive and kind, compassionate and domineering, optimistic and depressive, gallant and tactless, persuasive and coarse, pragmatic and idealistic (Byrd 1987, ix–xiv et passim).

DWIGHT D. EISENHOWER, 1890–1969

Dwight D. Eisenhower experienced a strong case of relative deprivation in childhood. Moreover, his military career was virtually fortuitous.

Eisenhower was the third of seven sons born to a "desperately poor" family in Denison, Texas (Lyon 1974, 39). Growing up in Abilene, Kansas, the Eisenhower children had a difficult time of it financially.

Lack of cash meant that the boys all had to wear hand-me-downs and in turn hand-me-downs meant jeers and ridicule from their schoolmates. When Dwight was the only one in his grade to wear overalls, when he or Edgar came to school wearing their mother's battered old button shoes, they could expect derision. They were ready for it, and met it with their fists. (Lyon 1974, 40)

By the time Eisenhower graduated from Abilene High School in 1909, he had developed a steely determination to succeed. But at what?

A friend told him about the service academies. At first, Eisenhower hesitated, since military service was against his family's pacifist conviction; "then he opted for the financial benefit over the religious or ethical injunction" (Lyon 1974, 41).

Eisenhower wrote Senator Joseph L. Bristow on August 20, 1910, requesting appointment to either Annapolis or West Point. He took his

West Point entrance examinations in March 1911 and entered the academy that June.

At West Point, Eisenhower had to prove himself all over again, particularly his physical superiority and his athletic prowess:

> Moreover, as had been the case in Abilene, he was encouraged at West Point to devote a great deal of his energy to athletics. This was so from his very first day when, because of his height, he was automatically assigned to F Company, traditionally the company of roughnecks, the jocks, the men who were proud of their muscles.... Eisenhower fitted well into F Company. He himself later recalled: "It would be difficult to over-emphasize the importance I attached to participation in sports." At his second summer camp he swaggered down F Company Street as he had swaggered across the railroad tracks on his way to high school in Abilene: self-confident, truculent, game for anything. (Lyon 1974, 44)

Unhappily, however, a serious knee injury in his second year at the academy finished his football career. He recalled: "I was almost despondent, and several times had to be prevented from resigning by the persuasive efforts of classmates. Life seemed to have little meaning; a need to excel was almost gone" (quoted in Lyon 1974, 44–45).

Athletics as an avenue of recognition having been closed, Eisenhower determined to become a full-fledged military officer.

> Professionalism, duty, honor, fealty to country, military tradition, political conservatism, belief in the inevitability of war, and the need for national armed might, and a readiness to fight and die—these were the chief tenets of the army officer's creed, these were the concepts which Eisenhower embraced in June 1915. (Lyon 1974, 48)

_____ Chapter 5

British Leaders

THOMAS FAIRFAX, 1612–1671

Thomas Fairfax "was a modest man, absolutely honest, and lacking personal ambition" (Wilson 1985, 189). At the same time, however, he was born to a military family and he experienced love deprivation.

Fairfax was born to a family whose traditions and ambitions were entirely military. John Milton characterized the family as "Fairfax whose name in arms through Europe sings" (quoted in Gibb 1938, title page). The entire family—father, grandfather, granduncle, uncles—impressed upon the boy the importance of following the family tradition and never wavering in loyalty to the king (Gibb 1938, 4, 5, 14; cf. Wilson 1985, 7).

Upon his mother's death in 1619, seven-year-old Thomas was sent to live with his strong-minded grandfather. "He seems to have spent a remarkably solemn boyhood. No stories of youthful pranks or escapades have come down to us. . . . [H]e spent too much of his time with adults, becoming melancholy, precocious, and introspective" (Gibb 1938, 4; cf. Wilson 1985, 12–13). At the same time, grandfather Thomas (first lord) continued to inculcate the family tradition:

> His grandfather doted upon him, impressing him at an early age with the glory of military life: "Tom, Tom, mind you the battle," he would say to him, "all the good I expect is from thee." He was not to be disappointed. (Gibb 1938, 4; cf. Wilson 1985, 7)

(The grandfather also instilled in Thomas an appreciation of literature and poetry, so that in his retirement—in the 1650s and the 1660s—Fairfax became a budding man of letters [Gibb 1938, Ch. 14].)

Having matriculated from Cambridge in 1629, Thomas's father Ferdinando sent him to the Netherlands—where he had contacts—to study military science. Upon returning to England in 1631, Fairfax joined the royal forces, rising steadily to become commander in chief of the Commonwealth Army during the English revolution of 1642–1649. He also played an important role in the formation of the New Model Army.

Two other aspects of Fairfax's life deserve mention. First, consistent with his passive personality, in 1637 he married Anne Vere, a strong-minded Presbyterian, who ran the Fairfax household (Gibb 1938, 14). Second, as a child, Fairfax may have experienced a degree of marginality: "Tall for his years, dark skinned and dark haired, he won, by reason of his swarthiness, the nickname 'Black Tom' " (Gibb 1938, 4).

JOHN CHURCHILL, 1650–1722

John Churchill's father, Winston, was an impoverished soldier and royalist politician. Experiencing serious relative deprivation in childhood, John joined the army as a means of social advancement.

As a youth, John was under the persistent influence of his father: "Winston himself instilled those principles of veneration for the institution of monarchy and the Protestant religion which his son never entirely forgot" (Ashly 1956, 14).

John's childhood was one of poverty and hardship. Born to a family of twelve siblings, of whom seven survived to adulthood, John's early life was characterized by persistent scarcity and want. As a result, he grew up to be avaricious:

> That he *was* mean and avaricious was widely accepted, both in England and on the Continent.... Branded deep in his character was the memory of his impoverished early life, and the insecurity and ignominy that went with it. (Barnett 1974, 178–79; cf. Ashly 1956, 14, 18; Churchill 1933–1938, I:ch. 1; emphasis in original)

Churchill's love of money reached a point when at times he contemplated accepting bribes (Bevan 1975, 290–91).

A man of "boundless ambition" (Ashly 1956, 81), John Churchill joined the army as a means of escaping his wretched family life. Having scored a series of military victories, he advanced steadily to become captain general of the army in 1702.

JOHN BURGOYNE, 1722–1792

John Burgoyne was born to a military family. He appears to have been an illegitimate child who sought military glory as a means of compensating for his marginality.

Burgoyne's father was an army captain who inculcated in the child the values of military life. From early on, Burgoyne developed a strong resolve to excel at the military profession.

Reinforcing Burgoyne's resolve was that he is widely reputed to have been the natural son of Lord Bingley, a close family friend (Hargrove 1983, 17–18; Howson 1979, 5–7; cf. Lewis 1973, 5). When Lord Bingley died in 1731, he left a significant portion of his estate to Burgoyne's mother (Anna Maria) and he forgave her husband (John Sr.) the debt he owed him (Howson 1979, 6).

Although illegitimacy was not uncommon in the eighteenth century— Burgoyne himself had four children out of wedlock—Burgoyne came to resent his father: "there is evidence to suggest that Burgoyne regarded his father with contempt, even hatred" (Howson 1979, 7). Having attended the prestigious Westminster School, Burgoyne ran away at age fifteen to join the Royal Army—possibly as a means of getting away from the father and the family.

Once in the military, Burgoyne gave full play to his ambition and vanity. "Burgoyne indeed desired to shine" (Hargrove 1983, 266). He became ready to sacrifice anyone or anything—including his principles—"to efface the shame of illegitimate birth by winning military glory" (Howson 1979, 5; cf. Glover 1976, 3–4).

A second way in which Burgoyne compensated was by cultivating social grace as a member of the lower aristocracy. He became charming, dashing, smooth, and polished; and he circulated in high society (Hargrove 1983, passim). For this he was nicknamed "Gentleman Johnny."

HORATIO NELSON, 1758–1805

Horatio Nelson experienced relative deprivation and love deprivation in childhood. His naval career was launched by his maternal uncle, Captain Maurice Suckling. His small stature created a sense of physical inferiority, for which he compensated by cultivating ambition, vanity, and glory.

Horatio's father, Rev. Edmund Nelson, was the rector of the local parish. Impecunious and frequently in poor health, he could barely provide for a family of eleven siblings, eight of whom survived to maturity. Horatio's early life was characterized by scarcity and want.

Horatio's mother, Catherine, died in 1767, creating a serious emotional void. "At the age of nine he suffered a traumatic experience which was to give him an Achilles heel: with the sudden death of his mother ... he was deprived of a woman's love and sympathy just when a boy needs them most" (Bennett 1972, 9). Accordingly, "With her death, his childhood ended" (Edinger and Neep 1931, 18). As a result, "By the time he was twelve, and perhaps because of the early removal of maternal influence, Horatio was already giving serious thought to launching out into the [military] world" (Grenfell 1950, 2).

Taking effective charge of Horatio's life from this point forward was his maternal uncle, Captain Suckling, later comptroller of the Royal Navy. In January 1771, when Horatio was only twelve, "Captain Suckling entered his nephew's name in the muster-book of his ship as a midshipman" (Pocock 1988, 7). Having joined the *Raisonnable*, "Suckling took a conscientious interest in the boy's professional upbringing" (Grenfell 1950, 4). Nelson rose to the rank of lieutenant by age nineteen and to captain by age twenty-one.

From early childhood—and throughout adulthood—Nelson experienced physical marginality. A biographer notes:

> He was never of a strong body; and the ague, which at that time was one of the most common diseases in England, had greatly reduced his strength; yet he had already given proof of that resolute heart and nobleness of mind, which, during his whole career of labor and glory, so eminently distinguished him. (Southey 1962, 2)

Alfred Thayer Mahan comments:

> He was himself, at various periods through his life, a great sufferer, and frequently an invalid; allusions to illness, often of a most prostrating type, and to his susceptibility to the influences of climate or weather, occur repeatedly and at brief intervals throughout his correspondence. (Mahan 1897, I:5)

A third biographer observes:

> Nelson was of small stature and had a plain lean face, bronzed by the sun. His whole personality left little impression of an imposing, manly, and heroic character. After having suffered the loss of an eye at Calvi [1794] and his right arm at Teneriffe [1797], he looked quite a cripple. In this insignificant-looking man, however, there dwelt a thoroughly well-balanced, vigorous mind, eager for great deeds. (Kircheisen 1931, 261; cf. Grenfell 1950, 2–3; Oman 1947, 10)

In addition to these handicaps, Nelson frequently experienced periods of depression and melancholy (Pocock 1988, passim; Bennett 1972, passim; Grenfell 1950, passim; Kircheisen 1931, passim). In a word, Nelson was a "neurotic" (Edinger and Neep 1931, 7).

Nelson compensated for his problems by cultivating vanity, ambition, and a thirst for glory. A biographer notes circumspectly:

> Yet he has remained a human being in memory and tradition: vivid, generous, and brave; sometimes vain; occasionally weak. . . . As [a friend] . . . Lord Minto

said of his contradictions: "He is in many points a great man; in others, a baby." (Pocock 1988, xii; cf. Bennett 1972, passim; Warner 1958, passim)

One of Nelson's "occasional weaknesses" was for the socialite Lady Emma Hamilton, with whom he carried on a torrid and scandalous liaison (while still married) between 1798 and 1800. Nelson met the Hamiltons in Naples, where Sir William was appointed British ambassador. Nelson became "a slave" to Lady Hamilton (Kircheisen 1931, ch. 4), who bore him a daughter in 1801, Nelson's only child. Lady Hamilton named the baby "Horatia."

Finally, we note an element of mysticism about Nelson: throughout his life he had a death wish. "He was firmly convinced that some day a bullet would strike him" (Kircheisen 1931, 274). As fate would have it, Nelson was killed by a French sharpshooter in the Battle of Trafalgar on October 21, 1805.

ARTHUR WELLESLEY (NÉ WESLEY), 1769–1852

(In 1798 Arthur Wesley returned to the original family name of Wellesley, possibly as a means of avoiding association with the evangelist John Wesley [1703–1791]. For convenience's sake, we shall refer to him as Wellesley throughout.)

Arthur Wellesley was born to a family with a long military tradition. He experienced a variety of difficulties and setbacks in childhood. His military career was initiated by his mother, Anne Wellesley.

Wellesley's father, Garrett, a member of the nobility and landed gentry, broke the family tradition of military service, most likely because in the eighteenth century the British Army stood in very low esteem and prestige (see, for example, Longford 1969, 18–19). Instead, he concentrated on cultivating his family and his estate.

Arthur Wellesley, an inactive, introspective, and lethargic youth (Longford 1969, 14; Aldington 1943, 2), faced intense competition from his brothers: "The childhood which has almost no annals is not necessarily happy. [Arthur's] grim reticence about his youth fits into the impression of a frustrated fourth child, conscious of latent powers but inhibited by two clever elder brothers and two promising younger ones" (Longford 1969, 13). Arthur compensated by emulating his artistic, music-loving father. In particular, he became adept at playing the violin.

Garrett Wellesley died in 1781, leaving the twelve-year-old Arthur at the mercy of a "cold and austere" mother (Fortescue 1925, 2): "The loss of his musical parent in May 1781 can scarcely have improved Arthur's prestige in the family. So far, his sole outstanding gift was intense love of music and skill in playing the violin. It was not a gift to be prized by his widowed mother" (Longford 1969, 15).

Arthur, who clearly preferred to follow in his father's footsteps and pursue a civilian life (Longford 1969, 18, 22), entered Eton in 1781. But he did not have the intellectual aptitude, he "learned absolutely nothing" (Aldington 1943, 22), and he abruptly ended his studies after three years.

Arthur's failure at Eton served only to reinforce Anne Wellesley's impression of him as "the dunce of the family" (Bryant 1971, 16). She variously described her son as "an ugly boy," "an awkward son," and "food for powder and nothing more" (quoted in Fortescue 1925, 5; Bryant 1971, 16; Aldington 1943, 26). Accordingly, in 1786, Anne dispatched Arthur to the Royal Academy of Equitation at Angers in Anjou, France, in order to prepare him for the British army.

During his year at Angers, Wellesley mastered fencing, horsemanship, swordplay, and mathematics—to say nothing of dancing, deportment, and graceful manners. Returning to England in 1787, a polished Wellesley made an "instant impression" on everyone—especially on his skeptical mother (Longford 1969, 21). "He is really a very charming young man," Anne noted, "never did I see such a change for the better in anyone" (quoted in Bryant 1971, 17).

Wellesley joined the British army in 1790. Three years later, in an act of supreme symbolism, he burned his violin (Longford 1969, 34). By 1814, he had become a field marshal.

There was an element of mysticism about Arthur Wellesley. He was fatalistic—a strong believer in the "finger of Providence" (see Longford 1969, 111, 330, 359, 427, 478, 490). He considered the British army "God Almighty's Army" (Longford 1969, 319).

CHARLES GEORGE GORDON, 1833–1885

Charles George Gordon was born to Henry William Gordon and Elizabeth Enderby, both of Scottish stock. Charles's dual absorptions were: (1) the military, which he inherited from his father; and (2) religion, which he internalized from his mother. At times, the two merged, and Charles George Gordon grew to become a soldier and a saint (cf. Garrett 1974, 14).

Charles was born in the garrison town of Woolwich, where his father eventually rose to become a lieutenant general. His great-grandfather and grandfather were also army officers. Of his four brothers, two joined the army; of his six sisters, three married army officers. The military was in Charles's blood.

Ambitious from an early age, Charles was determined to follow in the footsteps of his father, grandfather, and great-grandfather. "As a child he owned hundreds of lead soldiers which he 'drilled' for long hours at a time" (Nutting 1966, 6).

Like most military children, Charles never had a settled home. He moved from town to town and from garrison to garrison as the father's assignments dictated. "The constant moving about had brought out all the [volatile] Gordon in him, and for years he was little more than an irresponsible madcap" (Hanson and Hanson 1954, 16).

Following elementary school, Charles entered the Woolwich Military Academy at age fifteen (1848). Thoroughly undisciplined, he was subject to frequent reprimands for unruly behavior. Nonetheless, he graduated in 1852.

As he matured, Gordon came to develop a passion for war. He found war nothing short of exhilarating. He said he preferred to live in war than in peace (Hanson and Hanson 1954, 25).

In 1855, Gordon was sent to Crimea to fight on the side of the allies:

[O]n the day when a bullet "passed an inch above my nut," he announced that he had begun to "enjoy the work amazingly"—the "work" being not the building of huts but incursions into the trenches. War he found "indescribably exciting." He was soon noticed; not so much for his courage (for he had no need of it) as for his absence of fear. (Hanson and Hanson 1954, 22)

While still in Crimea:

The allies . . . opened a bombardment, and Gordon, running up the heights, saw "a splendid sight" that stirred him to the soul . . . "the whole town in flames, and every now and then a terrific explosion. The rising sun shining on the scene of destruction produced a beautiful effect." Then the retreating Russians blew up Fort Paul, and that too was "a beautiful sight." (Hanson and Hanson 1954, 25)

In 1860, England declared war on China. Gordon volunteered for active service, considering the experience as "the amusement" (quoted in Hanson and Hanson 1954, 30). Having spent five years in China, and having distinguished himself in the Taiping Rebellion (1850–1864), he earned the nickname "Chinese Gordon."

Gordon lived for war and died in war. While governor-general of Sudan, he was killed in action on the steps of the governor-general's palace in Khartoum on January 26, 1885.

Gordon's passion for war was mirrored in his passion for religion, which he absorbed from his mother, a "staunch evangelical" (Hanson and Hanson 1954, 16). "By his fond mother the boy had been taught to read and to believe implicitly in the literal truth of the Bible" (Hanson and Hanson 1954, 19). He came to accept that the Scriptures were "the direct voice of God" (Elton 1954, 299). Under the influence of his mother and the local ministers, "Gordon's spiritual life flourished. He was told that the world was rotten, the body the seat of evil, and that in the cultivation of the immortal soul alone rested man's hope and his justification" (Hanson and Hanson 1954, 19).

The mother's religious teachings were reinforced at Woolwich Military Academy, where Gordon formed a friendship with "a Captain Drew, a deeply religious man who was also something of an evangelist. This association was to form a permanent influence on Gordon's life and thinking" (Nutting 1966, 8; cf. Trench 1978, 15). The military and the religious, one might say, came to merge.

Gordon "did not identify himself with any particular church, but went here and there where true evangelism was preached" (Kübler 1912, 45). He also spent much time printing and distributing religious tracts.

As he matured, Gordon became a mystic and a fatalist—some say a fanatic:

Gordon had an unshaken belief in God and His fatherly direction in everything. On this account, he has often been called a fatalist and sometimes a Mohammedan. He was a fatalist only in the sense that whatever happens, great or small, is wisely ordered by the Divine Being for our good. He did not believe, however, that one should sit still and fold his hands in his lap. He must do his work faithfully, industriously, and prayerfully.... He also believed that evil was permitted and regulated by God, and this conviction gave him all the more strength and energy in accomplishing whatever he had to do....

The main principle actuating Gordon was the real presence of God in the believer. He believed this in the truth of the precept "God in us" in common with ... mystics. ...

This belief in the dwelling of God made Gordon unspeakably happy and raised him so far above this world that he was not affected by its praise or its blame. (Kübler 1912, 82–84; cf. Hanson and Hanson 1954, 23, for characterization of Gordon as a fanatic)

Accordingly, Gordon lived an almost ascetic life. He never married, he shunned parties and dances, he disliked social life.

We close by noting "Gordon's own repeated statements that in most of his life he was motivated by a death-wish" (Nutting 1966, 318). Whether this death-wish accounted for his courage and bravery as a soldier, or whether it emanated from his religiosity, is open to speculation.

HORATIO HERBERT KITCHENER, 1850–1916

Horatio Kitchener was born to a military family. He encountered prejudice while going to school in Switzerland. He experienced love deprivation as a result of losing his mother in childhood.

Kitchener's father, Henry, was a colonel in the British army. Henry Kitchener taught his children "self-discipline and mutual discipline"

(Warner 1985, 10). The Colonel was "an eccentric martinet who ran his home as far as possible on army lines.... Colonel Kitchener, who was proud, fearless and independent, brought up his children to regard themselves members of a master race" (Magnus 1959, 4).

It is not surprising that of Colonel Kitchener's four sons, three went into the army. This included Horatio, who as a child was aloof, withdrawn, and reserved (Magnus 1959, 5). A biographer notes: "shy, sensitive, backward and unathletic as [Horatio] Kitchener was, he never abandoned his ambition to obtain a commission in the Army" (Warner 1985, 12).

In 1864, the Kitchener family moved from Ireland to Switzerland in order to seek medical treatment for the ailing mother, Frances Ann. Kitchener and his three brothers went to a boarding school, where they experienced hostility and prejudice:

Kitchener had been very greatly attached to his mother, who thought he was perhaps too sensitive for his own good, and now at the age of fourteen he found himself in a hostile environment where he and his brothers were despised for their almost total ignorance of subjects with which their contemporaries were familiar. Physically they could look after themselves, but there was no avoiding the mockery and contempt which boys of no greater educational depth themselves will enthusiastically lavish on companions of lesser attainment. (Warner 1985, 11–12; cf. Magnus 1959, 7)

This experience made Horatio realize that he could succeed only "by his own efforts" (Warner 1985, 12).

Kitchener's beloved mother died later that year, setting in motion a chain of love deprivation. Returning to England in 1867, Horatio entered the Royal Military Academy at Woolwich a year later, graduating in 1871.

An intensely ambitious man, Kitchener's entire adult life—between 1871 and his death in 1916—was given over to a constant series of military assignments throughout Europe, Africa, the Middle East, and India. A bachelor without family ties, Kitchener developed an intense passion for the army (Magnus 1959, 10). He saw every military victory as "an ambition achieved" (Warner 1985, 4).

BERNARD LAW MONTGOMERY, 1887–1976

Bernard Law Montgomery was born to Henry Montgomery, a clergyman, and Maud Montgomery, a tyrannical housewife with whom Bernard was in constant conflict. As the fourth child and third son in a family of nine children, Bernard felt alone, neglected, and cast aside. From early on, he felt the need for self-direction and self-control. In particular, he developed a steely determination to join the military as a means of escaping a wretched home life.

A strong-willed woman, Maud Montgomery "began to use her domestic hegemony as a penal lever on her hardworking, missionary spouse. Within a very short time he was emasculated of all domestic authority. Maud ruled; and her rule ... was sometimes despotic and often blind" (Hamilton 1981, 4–5). Montgomery recalled:

> Certainly I can say that my own childhood was unhappy. This was due to the clash of wills between my mother and myself. My early life was a series of fierce battles, from which my mother invariably emerged the victor. If I could not be seen anywhere, she would say—"Go and find out what Bernard is doing and tell him to stop it." But the constant defeats and the beatings with a cane, and these were frequent, in no way deterred me.... I never lied about my misdeed. I took my punishment. (Montgomery 1958, 17)

A biographer notes that of the Montgomery children,

> Bernard was undoubtedly the black sheep. Whether he would have risen to the heights he did had he not resented and fought against this loveless tyranny one cannot say; but certainly his awkward, cussed, and single-minded character was molded in the struggle of wills between them. Though he came to respect his mother, he never forgave her.... [Of the Montgomery children] only Bernard challenged her; and the conflict of their personalities lasted until her death in 1949. He did not attend her funeral. (Hamilton 1981, 5)

It was not surprising, as Montgomery noted later, that "under such conditions all my childish affection and love was given to my father. I worshiped him. He was always a friend. If ever there was a saint on this earth, it was my father. He got bullied a good deal by my mother and she could always make him do what she wanted" (Montgomery 1958, 19).

In the thick of this family conflict, Bernard sought his freedom, "not in emigration or early marriage [as his siblings had done], but in the sword. There, in the company of men in the life-or-death arena of war, he could both vent his frustrated emotions and earn the respect he courted" (Hamilton 1981, 6).

In 1889, Henry Montgomery was appointed Bishop of Tasmania, where the family stayed until 1901. During those years, Bernard was educated by English tutors. Having returned to England, the family enrolled Bernard in St. Paul's School in London in January 1902. Not being intellectually oriented, on the first day of school Bernard joined St. Paul's lowest echelon, the Army Class, and announced his decision to his parents that evening: "I want to be a soldier" (quoted in Moorehead 1946, 34). The parents were disappointed (Henry wanted Bernard to become a clergyman), but they did not put up a fight (Montgomery 1958, 22; Hamilton 1981, 42–43).

It would appear that at St. Paul's, Bernard took the easy way out. He acknowledged later that "In those days the Army did not attract the best brains in the country" (Montgomery 1958, 23). Therefore, he could keep company of the intellectually mediocre: "Bernard could dawdle and dream his way through school years, concentrating only on the new sports which St. Paul's offered: cricket and rugby football. Within three years he was captaining both teams" (Hamilton 1981, 43).

Success in sports, in turn, intensified Montgomery's interest in the military: "For the first time in my life leadership and authority came my way; both were eagerly seized and both were exercised in accordance with my own limited ideas, and possibly badly. For the first time I could plan my own battles (on the football field) and there were some fierce contests" (Montgomery 1958, 20).

Having completed his studies at St. Paul's in 1906, Montgomery took the entrance examinations for the Royal Military Academy in the fall of that year and entered Sandhurst in January 1907, graduating a year and a half later. By that time, "a very important principle had just begun to penetrate my brain. That was that life is a stern struggle, and a boy has to be able to stand up to the buffeting and the setbacks" (Montgomery 1958, 21–22).

As he matured, the self-confidence to which Montgomery refers throughout his memoirs (Montgomery 1958, passim)—and to which we have alluded above—took the form of monumental vanity. Even as a young man "he evidently became a monstrous bully, torturing recalcitrant colleagues and imposing his own tyrannical will, as his mother did within the home" (Hamilton 1981, 39). As an adult, his egomania became legendary. He carved a reputation for routinely and ruthlessly denigrating his colleagues (Hamilton 1981, 1983, 1987, passim).

Finally, we note an element of mysticism about Montgomery. He titled the epilogue to his book on military leadership (Montgomery 1961) "In My Garden," where, falling asleep, he carried on a conversation with the ghost of his beloved father.

LOUIS MOUNTBATTEN (NÉ BATTENBERG), 1900–1979

Louis Mountbatten emulated his father, Prince Louis of Battenberg, throughout his life. In addition, ambitious from childhood, he sought to avenge and vindicate his fallen father.

Born in Graz, Austria, in 1854, from childhood Prince Louis "passionately wanted to be a sailor" (quoted in Terraine 1968, 17). Since there was no Austrian or German navy at that time, he sought an alternative route in England.

A grandson of Queen Victoria and Prince Albert, at age eight Prince Louis was taken to live with the royal family in England. "Young Louis fell in love with England, and above all fell in love with the British navy" (Hough 1981, 11). Having joined the Royal Navy in 1868, he was ridiculed for his strong German accent. Through hard work, though, he rose in the ranks to become a captain in 1900. He also strove to become "more English than the English" (quoted in Hough 1981, 12). By 1914, Prince Louis had risen to the rank of the First Sea Lord.[1]

This was the family environment into which Louis ("Dickie," as he was nicknamed) was born on July 25, 1900. From the earliest childhood, he was thoroughly socialized in the way of the British Navy. Beginning at age four, his parents dressed him in naval uniforms. Dickie observed as a youngster: "It never at any remote moment entered my mind—it never even occurred to me—I had no other plans whatsoever—than go into the Navy" (quoted in Hough 1981, 20). This sentiment was thoroughly shared by the parents in the close-knit family (Ziegler 1985, 30; Hough 1981, 10; Terraine 1968, 4). Accordingly, Dickie joined the Royal Navy in 1913, at age thirteen. He attended the Osborne and Dartmouth naval academies, and for a time studied at Cambridge.

In 1913 disaster befell the Battenberg family. Following British naval setbacks at the hand of Germany at the beginning of World War I, Prince Louis was wrongly accused of mismanagement and even of spying for Germany. As a result, he was forced to resign as the First Sea Lord. Given the anti-German hysteria that swept Britain at that time, in 1917 Prince Louis of Battenberg changed the family name to Mountbatten.

The father's forced resignation struck Dickie with extreme force. "Shocked," "stunned," and "outraged" (Hough 1981, 30–31; Terraine 1968, 14–15), he redoubled his efforts not only to follow in the father's footsteps, but also to assume the position of First Sea Lord. He noted: "Ever since that disgraceful episode, I have determined to get to the top and vindicate his memory. Nothing and no one; I repeat, nothing and no man, will ever be allowed to stand in my way" (quoted in Ziegler 1985, 334).

Mountbatten finally realized his ambition forty-one years after his father's forced resignation, being named First Sea Lord in 1955. In addition, a year later, he was promoted to the rank of fleet admiral. A great deal of hard work at last paid off.

NOTE

1. "The First Lord was the civilian head of the admiralty, a political post appointed by the Prime Minister. The term is now obsolete. The First Sea Lord, working under the First Lord at that time, was the professional head of the Royal Navy" (Hough 1981, 21n).

_____ Chapter 6

French Leaders

NAPOLEON BONAPARTE (NÉ NAPOLEONE DI BUONAPARTE), 1769–1821

Napoleon Bonaparte reflects an impressive range of the psychological–situational dynamics we associate with military leaders.

Napoleone di Buonaparte was born to a family of lesser and impecunious Italian (Tuscan) aristocracy on the island of Corsica on August 15, 1769. He adopted the more familiar name upon becoming a general in the French army in 1796, at age twenty-six. For convenience's sake, we shall use Napoleon Bonaparte throughout.

Historically, Corsica had been subject to invasions by Phoenicians, Etruscans, Romans, Byzantines, Lombards, Berbers, and others. In the 1760s and 1770s, Corsica was in the midst of a great nationalist struggle. In particular, Genoa and France were in competition over Corsica, the latter finally annexing the island in 1768 and making it formally a part of France two years later.

Napoleon was born during the Corsican nationalist conflict. A strong admirer of Pasquale Paoli, the Corsican nationalist leader, in his youth Napoleon was consumed by Corsican nationalism, seeking independence from France (Watson 1902, 1–46).

Within a decade or two, Napoleon experienced a dramatic turnabout by becoming a Francophile and a French nationalist. The first turning point came in 1789 as Napoleon correctly perceived opportunities for personal advancement in the French Revolution. By some accounts, Napoleon became an admirer of Robespierre and Marat; for a time, he was a Jacobin (Watson 1902, 82, 90; Ludwig 1924, 16). A second turning point came in the French victory over England at Toulon in 1793. A third turning point was Napoleon's assumption of the role of First Consul in 1799. A final turning point was his self-coronation as emperor in 1804.

Napoleon remained a French nationalist until his last days. To this effect, one finds a variety of entries in his diary: (1) March 6, 1815: "I live only for the honor and for the happiness of France" (Bonaparte 1910, 448); (2) March 26, 1815: "Henceforth the happiness and consolidation of the French Empire will be the subject of all my thoughts" (Bonaparte 1910, 450); (3) June 1, 1815: "In prosperity, in adversity, on the battlefield, in Council, on the throne, in exile, France has been the one and only object of my thoughts and my deeds.... Frenchmen, my will and my duties are those of the French people; my honor, my glory, my happiness, can be none other than the honor, the glory, and the happiness of France" (Bonaparte 1910, 454). (As these pronouncements were made immediately after Napoleon's exile on Elba and only months before his exile on St. Helena, their reliability and sincerity may be open to question.)

Bonaparte showed a precocious interest in military matters. This interest was actively cultivated and advanced by his father, Carlo, a Corsican nationalist turned Francophile—so much so that he changed his first name to Charles. Even as a child, Napoleon spent a great deal of time playing with toy soldiers and arranging them in parades (Barnett 1978, 16; Kircheisen 1932, 8). By the time Napoleon was seven, Carlo had decided that his son should prepare for a military academy (Cronin 1972, 27). When Napoleon was nine (in 1779), through a Corsican contact, Carlo registered Napoleon in elementary school at Autun, France, where he stayed for only three months. As per French policy, the authorities selected the promising youngster to attend the military academy at Brienne, where he trained for five years. Having done well at Brienne, Napoleon was selected in 1784 to attend the higher military academy (École Militaire) in Paris. A year later, he passed his examinations, becoming a second lieutenant in the artillery. In short, Napoleon's early military career was shaped by an inner drive, an ambitious father, and a fortuitous French policy.

In addition to his Corsican background, Napoleon remained marginal in French society in a variety of ways:

> We shut our eyes.... We see a lean, sallow, awkward, stunted lad step forth from the door of the old house and go forth into the world, with no money in his pocket, and no powerful friends to lift him over the rough places. He is only nine years old when he leaves home, and we see him weep bitterly as he bids his mother good-by. We see him at school in France, isolated, wretched, unable at first to speak the language, fiercely resenting the slights upon his poverty, his ignorance, his family, his country—suffering but never subdued. (Watson 1902, 17; cf. Kircheisen 1932, 9; Castelot 1971, 5)

In particular, Brienne posed most trying times for the young Napoleon:

The pupils were mainly aristocratic French scions of the privileged nobility, proud, idle, extravagant, vicious. Most of these young men looked down upon Napoleon with scorn. In him met almost every element necessary to stir their dislike, provoke their ridicule, or excite their anger. In person he was pitifully thin and short, with lank hair and awkward manners; his speech was broken French, mispronounced and ungrammatical; it was obvious that he was poor; he was a Corsican; and instead of being humble and submissive, he was proud and defiant. During the five years Napoleon spent here he was isolated, moody, tortured by his discontent, and the cruelty of his position. (Watson 1902, 26; cf. Kircheisen 1932, 10–11)

Napoleon lashed back: "I am tired of explaining my poverty; of having to endure mockery of these foreign boys, whose only superiority is in respect of money, for in nobility of feeling they are far beneath me. Must I really humble myself before these purse-proud fellows?" (quoted in Ludwig 1953, 5).

The situation did not improve much at École Militaire in Paris: discrimination and taunting continued unabated. And so it remained to the end. Napoleon never developed "grace of speech and manner"; there was "nothing imposing in his figure and bearing"; he remained coarse and unkempt (Watson 1902, 100, 129). Early in his career, soldiers derisively called him the "Little Corporal." Even at the height of his power, Napoleon was never fully accepted by French aristocracy, who variously described him as "the Corsican upstart" and "the Corsican ogre" (Watson 1902, 414, 466). In short, "With his elevation to empire, Napoleon became more stately, reserved, dignified, and imposing; but perfect ease and repose of manner he never acquired. The indolent, calm, and studied air of languor and fatigue which, according to a well-known standard, constitutes good breeding, he did not have" (Watson 1902, 390).

Not surprisingly, Napoleon compensated for his marginality by becoming distant, aloof, arrogant, and imperious. What can one say of Napoleon's vanity, egomania, and self-conceit that has not been said before? Shall we note his thirst for power: "Power is my mistress. I have worked too hard at her conquest to allow anyone to take her away from me or even covet her" (quoted in Herold 1955, 257). Shall we observe his self-concept: "Men who are truly great are like meteors: they shine and consume themselves, that they may lighten the darkness of the earth" (quoted in Ludwig 1953, 20; cf. Castelot 1971, 403). Shall we note his self-coronation as emperor? Shall we observe the style of imperial proclamations after 1804: "Napoleon by the grace of God Emperor" (quoted in Watson 1902, 414). Shall we note his designs to conquer Italy,

Austria, Germany, Spain, Portugal, England, Poland, Russia, Egypt, India, Turkey, and Persia? While in exile on St. Helena, he entered in his diary for March 3, 1816:

The obstacles before which I failed did not proceed from men but from the elements: in the south it was the sea [that] destroyed me; in the north it was the fire of Moscow and the ice of winter; so there it is, water, air, fire, all nature and nothing but nature; these were the opponents of a universal regeneration commanded by Nature herself! The problems of Nature are insoluble! (Bonaparte 1910, 471–72)

Finally, we note an element of mysticism about Napoleon: "He was not free from superstition. What people called 'omens' made an impression upon him. He sometimes made the sign of the cross, as though to ward off impending evil" (Watson 1902, 406–407). He wrote in his diary for July 30, 1800: "In warfare every opportunity must be seized; for fortune is a woman: If you miss her today, you need not expect to find her tomorrow" (Bonaparte 1910, 143). In 1815, just before the battle of Waterloo, he observed: "I felt that Fortune was abandoning me. I no longer had the feeling that I was sure to succeed. . . . When a man does not act boldly, he never does anything at just the right moment; and a man does not act boldly unless he is convinced that luck is on his side" (quoted in Castelot 1971, 535).

GILBERT LAFAYETTE, 1757–1834

Gilbert Lafayette was born into a military family. Having experienced serious love deprivation in childhood and youth, he set out in search of a surrogate father, whom he eventually found in George Washington. In addition, he struggled with an inferiority complex in his youth. Lafayette had no formal military training; his military career was entirely fortuitous.

Marquis de Lafayette, a colonel, was killed at the Battle of Minden in 1759, uprooting his two-year-old son altogether. A year later, Madame de Lafayette left the family village of Auvergne to live with her parents in Paris, leaving behind her young son. Gilbert was looked after principally by his maternal grandmother, Madame du Motier.

In 1768 Madame de Lafayette sent for her eleven-year-old son, plunging Gilbert into a state of social and cultural shock upon reaching Paris. Almost immediately, Gilbert was sent to College du Plessis, where he felt lonely and isolated. By all accounts, Gilbert was awkward as a youth, not feeling at home in the social atmosphere and the salons of Paris. As a biographer put it, "To Gilbert, this extraordinarily privileged world was simply his private hell" (Bernier 1983, 10).

In 1770 Madame de Lafayette died unexpectedly at age thirty-two. Although the thirteen-year-old Gilbert had the care and attention of his grandmother and other relatives, "a vital support had been withdrawn. Though he did have guardians, no one was in charge of his life any more" (Bernier 1983, 10).

In 1771 Gilbert decided to embark on his military career. An influential family member secured him a commission as a lieutenant in the Black Musketeers, "an elite corps of which his great-grandfather had been the commander" (Bernier 1983, 11). By 1774 (at age seventeen), through family influence, Lafayette was promoted to captain, a rank which he "in no way deserved" (Bernier 1983, 16).

The year 1774 brought another major event in Lafayette's life: he married Adrienne de Noailles, daughter of one of the most aristocratic families in Paris, with close connections to the court of Louis XVI. "The new social advantages only exacerbated Lafayette's feeling of inferiority. Now, more than ever, life was thoroughly unpleasant. First, he dreaded his father-in-law who, like the accomplished courtier he was, valued polish, wit, and grace above all else" (Bernier 1983, 17). Moreover, he suffered by comparison with the other son-in-law, who possessed all the desired qualities. As a result, Duke de Noailles (the father-in-law) routinely ridiculed Lafayette.

Nonetheless, things continued to go well for Lafayette. "The future was secure: Gilbert knew that he would rise in the army through his connections at court, that Adrienne would become a lady-in-waiting to the Queen. With luck and the Noailles influence, he could look forward eventually to being created a duke" (Bernier 1983, 18).

But first, Lafayette had to prove himself to be court-worthy. So, he tried to imitate the aristocracy as best he could—dancing, gambling, horseracing, and the like. "In spite of his best efforts, however, Lafayette never felt he really belonged" (Bernier 1983, 18). A contemporary described him as "awkward and gauche"; he was "dreadfully uncomfortable at Versailles" (Bernier 1983, 18, 21).

Given the opportunity suggested by a superior officer, Lafayette welcomed the chance to join the American Revolution. His personal motive was escaping the miseries of the family and the court. The larger motive was a chance to take revenge on Paris. "With just a little luck, the young man would return to France a hero. That would show the carping... [adversaries and detractors] and the world in general just what a mistake they had made" (Bernier 1983, 25).

Meanwhile, in June 1776, a new reform-minded war minister placed Lafayette and a number of other young officers on the inactive list. For all practical purposes, Lafayette's military career came to an end. "Gilbert was a hopeless failure" (Bernier 1983, 25). In the fall of 1776, Lafayette decided to volunteer as an officer in the American army.

In April 1777, along with a few other officers, Lafayette bought a ship and set sail for America in search of glory, honor, and heroism. In June 1777, he landed at Georgetown, South Carolina, where he was welcomed with open arms. He traveled widely, meeting many dignitaries and gaining widespread respect to replace the scorn he had experienced in Paris. In July 1777, without any military experience whatever, the Continental Congress appointed him a major general. Lafayette offered to serve as a volunteer without pay or pension as a means of (1) enhancing his credibility, and (2) assuring his freedom to leave whenever he pleased.

Shortly thereafter, Lafayette met his commanding officer, George Washington. It was a mutual admiration society. "For Lafayette it was reverence at first sight" (Bernier 1983, 45). Washington, who was childless, reciprocated Lafayette's admiration and affection. The two became virtually father and son. Washington invited Lafayette to come live in his house—a singular honor. In short, in meeting Washington, Lafayette "had at last found the father figure he had been looking for" (Bernier 1983, 46; cf. Loth 1951, chs. 2 and 7).

Lafayette fought with bravery, distinction, and courage, gaining the respect and admiration of Washington and other generals. Lafayette knew nothing about the military, but because he demonstrated enthusiasm, modesty, and willingness to undertake any assignment,

> he was immediately accepted by his new peers; and because, for the first time since his childhood, he found himself surrounded by approval, he developed "those engaging manners" Benjamin Franklin was soon to note. By the end of August [1777], the awkward, gawky young man was nothing but a memory. (Bernier 1983, 48)

As a result, "what seemed lacking in Versailles became dazzling on the other side of the Atlantic" (Bernier 1983, 53). Lafayette reinforced all this by talking and acting as if he had been at the center of social and political power in Paris and Versailles.

In December 1777, with Washington's intercession, the Continental Congress appointed Lafayette to the command of a division of the Continental Army, something Lafayette had avidly and persistently sought. In February 1778, he was appointed general of the Northern Army.

In the fall of 1778, with the American Revolution slowing down because of the approaching winter, and with rumors of a French invasion of England, Lafayette decided to return to France. The American leaders issued proclamations in his honor. Lafayette set sail in January 1779, reaching Paris in February—to great acclaim as a hero and a patriot. He was "the hero of the hour"; he was "the toast of Paris"; he was "someone to be reckoned with at Versailles"; he received "extraordinary and universal acclaim" (Bernier 1983, 82, 85). He was promoted to the rank of colonel in the French army.

In April 1780, Louis XVI and his premier Comte de Maurepas sent Lafayette back to America as a representative of France. Lafayette continued to fight with enthusiasm up through the Battle of Yorktown (October 19, 1781), where General Charles Cornwallis surrendered to General George Washington. In January 1782, Lafayette returned to Paris, once again to great popular acclaim.

In 1780, Lafayette named his only son George Washington; two years later, he named his daughter "Virginie" in honor of Virginia. Washington's status as a father figure and a hero persisted throughout Lafayette's life.

Given Lafayette's accomplishments, we should not fault him for his growing vanity, his "inflated ego," his "endless appetite for flattery," his "thirst for popularity" (Bernier 1983, 329, 330, 331). Perhaps he earned all this.

(Lafayette's involvement in French politics and revolution in the 1780s and beyond shall not concern us for the purposes of this work.)

FERDINAND FOCH, 1851–1929

Ferdinand Foch was born into a military family with a tradition of profound reverence and veneration for Napoleon. Ferdinand's maternal grandfather had been an officer in Napoleon's Italian Army, and he had been decorated by Napoleon (Laughlin 1919, 17; Liddell-Hart 1928, 152). His father-in-law had also served under Napoleon. His father's Christian name was Napoleon, and the three Foch brothers were known as "little Napoleons" (Liddell-Hart 1931, 6). "The young Ferdinand was brought up on a sustained diet of Napoleonic fervor" (Marshall-Cornwall 1972, 1; cf. Laughlin 1919, 29). He was "reared in an atmosphere of Napoleon-worship" (Aston 1929, 12).

It was natural, then, that from very early on, Foch would determine to emulate Napoleon. By age ten, "He knew most of Napoleon's campaigns by heart" (Aston 1929, 17). From age eighteen onwards, "Foch's ambition was to gain a Commission in the Artillery, thus following in the footsteps of the great Napoleon" (Marshall-Cornwall 1972, 2; cf. ibid., 6). Having attended the École Polytechnique, and having enlisted in the Franco-Prussian War of 1870, he was commissioned in 1873, becoming an artillery officer a year later. Having attended the École Supérieure de Guerre in 1885, Foch's career skyrocketed—he become a general in 1914 and a Marshal of France four years later.

The Franco-Prussian War had a profound influence on Foch and other young recruits: "The indignities suffered at the hands of the occupying troops must have rankled in the breasts of the young students who had been reared in the afterglow of Napoleonic triumphs.... The idea of *la ravanche* now took possession of Foch's mind" (Marshall-Cornwall 1972, 3; cf. Aston 1929, 31–32). Foch became intensely nationalistic. He "dreamed constantly of the future glory of France" (Aston 1929, 45).

With the reverence for Napoleon and the glorification of France came an element of vanity and egotism. According to Foch: "The will to conquer sweeps all before it. There is a psychological phenomenon in great battles which explains and determines their results.... What compels victory is, above all else, the conduct of the commander" (quoted in Liddell-Hart 1928, 154–55). And again: "Great results in war are due to the commander"; Foch was fond of quoting Napoleon's aphorism: "Caesar, and not the Roman legions, conquered Gaul, and Rome trembled before Hannibal, not because of the Carthaginian soldiery" (quoted in Liddell-Hart 1931, 25).

Foch's military ambition, nationalism, and vanity were undergirded by religion. Foch's parents were devout Catholics who imparted that tradition to Ferdinand, which was reinforced in Jesuit school. "Religion formed his moral basis. It gave him the calmness of soul, the equilibrium needed by mankind. That was its great function. Duty, sacrifices, the complete subservience of physical instinct to spiritual aspirations, were made easier by religion" (Recouly 1929, 302). Accordingly, Foch became dedicated to hard work, living an almost ascetic life. "The essence of Foch's teaching was thus a powerful reassertion of the moral factor, particularly in the leader" (Liddell-Hart 1928, 156; cf. Liddell-Hart 1931, 5–6; Aston 1929, 7–8).

PHILIPPE PÉTAIN, 1856–1951

Philippe Pétain came from a family of solid Artois peasant stock, with no military background. He was inspired to join the army by the stories and anecdotes of a grand-uncle who had served under Napoleon.

Philippe's mother died in 1857, when the child was one-and-a-half years old. Omer Venant Pétain remarried soon thereafter. Unhappily, however, Philippe's stepmother was indifferent to—and hostile toward—Philippe and his three older sisters. (In time, the new Madame Pétain produced four more siblings.) As a result, Philippe lived with his grandparents—in effect, motherless, fatherless, and quite possibly in the throes of love deprivation.

Philippe attended a variety of religious (Dominican) and public schools, of which the most important was Saint Omer:

> Saint Omer was a garrison town, headquarters of a battalion of light infantry (one day he would command it). He had come to know most of the lieutenants by sight; in some of them he discovered models for his own career. "From that

time on my choice was made: I would be a lieutenant in the light infantry." Reports from the battlefields of Franco-Prussian War [when Philippe was fourteen years old] stirred the pupils, who spent their recreation periods in military drill. "I made myself a captain with the tacit agreement of my soldiers, and during a period of several months the school's yard echoed the sounds of military orders, while true military operations were saved for our outings, usually taking the form of an attack and defense of a fortress." (Lottman 1985, 22)

Pétain entered the military academy at Saint Cyr in 1876, at age twenty, bent on revenge against the Germans/Prussians. He graduated a second lieutenant two years later:

Saint Cyr was perhaps one of the strongest influences in Pétain's life. He learned there not only a respect for military virtues but also the military contempt for politicians and civilians, the mistrust for Republican education and the doctrines of the left, and the respect for tradition, which were to mark his later thought. He also developed that straight-faced irony which was to be one of his most effective methods of communication. (Griffiths 1970, xv)

Pétain attended the École Supérieure de Guerre in 1888 and was promoted to the rank of major two years later. For the following decade and a half his career became stagnant.

In 1914, at age fifty-eight, Pétain was a colonel resigned to his fate. World War I transformed a routine career destined for retirement in obscurity. The Battle of Verdun (1916) turned Pétain into a national hero. Two years later he was promoted to Marshal of France.

"Now Pétain was *persona grata* in the Paris salons," where he met and conversed with members of the French aristocracy; everywhere he experienced "the adulation of others" (Griffiths 1970, 94, 212). As a result, his vanity blossomed, and he became convinced of his own superiority. In fact, "Convinced that he alone as an 'incarnation' of France, could save his country, he acted on this basis" (Griffiths 1970, xi).

(Pétain's later involvement in French politics, including his role in the Vichy government, shall not detain us for the purposes of this work.)

CHARLES DE GAULLE, 1890–1970

Charles de Gaulle was born to a military family. From early on, he developed a determination to aggrandize and glorify France. He saw himself as the sole agent of that aggrandizement/glorification.

De Gaulle came from a military family tradition going back to the fifteenth century (Hatch 1960, ch. 2). An early ancestor, Jehan de Gaulle, fought the British at the Battle of Agincourt in 1415. Charles's father received a military education and fought in the Franco-Prussian War of

1870. Having obtained a doctorate, he subsequently taught literature and philosophy at a Jesuit college.

The de Gaulle family was close-knit:

> De Gaulle was close to both of his parents all his life. From his mother came the passionate, sensitive side of his nature, as well as his mask of reserve. His father, a very tall man who passed his height on to his four sons, gave him his intelligence, his deep and systematic way of thinking, his sense of history, his application and self-discipline, and his great strength of character. (Cook 1983, 27)

De Gaulle's parents were intensely religious and patriotic. They had "two moving passions in life, piety and patriotism. . . . Their children were imbued with a living faith in God and so deep a love for France that it became for all of them the central theme of their existence" (Hatch 1960, 21; cf. Tournoux 1966, 3, 7). In particular, the whole family was obsessed with the idea of *la ravanche* against Germany as a result of the defeat in the Franco-Prussian War (Tournoux 1966, ch. 2).

Charles became a nationalist to the core from a very young age, obsessed with the grandeur of France. At ten, he played with toy soldiers with his three brothers on the map of Europe, assigning them their roles: "You, Xavier, are Austria. Jacques is Prussia, Pierre is Italy. Of course, I am France" (quoted in Hatch 1960, 15; cf. Tournoux 1966, 7). De Gaulle confided later: "The emotional side of me tends to imagine France, like the princess in the fairy stories or the Madonna in the frescoes, as dedicated to an exalted and exceptional destiny" (quoted in Cook 1983, frontispiece; cf. Hatch 1960, 16).

The de Gaulle family spent summers in their home in the country:

> The summer he was thirteen he organized the neighboring farmer boys into a troop of scouts . . . and took them on a camping trip which was, in fact, a campaign. Though most of the boys were older than de Gaulle there was no question of who was in command, a thing made easier by the fact that he was already taller than most eighteen-year-olds. . . .
>
> For amusement they divided into opposing armies and refought famous battles. Those that France had actually won, followed the facts; but historical accuracy suffered severely when a defeat like Agincourt was replayed, for de Gaulle, always commanding for France, revised the faulty strategy and won. (Hatch 1960, 22–23)

By age fifteen, Charles already knew himself as "General de Gaulle"; two years later he came to see himself as "General and Commander-in-Chief" (Lacouture 1990, 3; Tournoux 1966, 16).

Having attended parochial and public schools, de Gaulle entered Saint Cyr in 1910, graduating two years later: "To his love for France was added a passion for the Army, for which he had, in truth, a vocation in the religious sense of the word" (Hatch 1960, 28).

While at Saint Cyr (as in earlier schooling), classmates made fun of de Gaulle for his gangling frame (six feet, five inches) and for his long, prominent nose. For his height, he was nicknamed *La Grande Asperge* (Grand Asparagus Stalk); for his nose, "Cyrano" (Cook 1983, 28; Hatch 1960, 29). This was the extent of marginality de Gaulle experienced in life.

What shall we say of de Gaulle's egotism and vanity? He was convinced from an early age that he was destined to render France singular service. He developed an unshakable vision of greatness—of himself and of France: in fact, he came to equate himself with France. After all, did "de Gaulle" not mean "of France"? (Lacouture 1990, 282 et passim; Cook 1983, 15–16 et passim; Hatch 1960, 15–16 et passim). No wonder that at least one biographer has characterized de Gaulle as "a great mystic" (Hatch 1960, 265, 266).

Jean Paul Sartre is supposed to have said that if he had a choice between de Gaulle and God, he would choose God, for He is more modest! (PBS program on "De Gaulle and France," November 30, 1992; rebroadcast June 1, 1994). De Gaulle himself acknowledged: "Every man of action has a strong dose of egotism, pride, hardness, and cunning. But all those things will be forgiven him, indeed, they will be regarded as high qualities, if he can make of them the means to achieve great ends" (de Gaulle 1960, 64).

The reference to the philosopher Sartre brings into focus another dimension of de Gaulle's personality—his early preoccupation with literature and philosophy. As a family tradition (recall that his father was a professor of humanities), de Gaulle read widely in literature, philosophy, and history: Shakespeare, Goethe, Nietzsche, Jules Verne, Henri Bergson, and others. Bergson was a family friend who visited occasionally. "The realism of Bergson's method of reasoning appealed to Charles's logical mind" (Hatch 1960, 23).

_____ Chapter 7

German Leaders

CARL VON CLAUSEWITZ, 1780–1831

Carl von Clausewitz was born to a Prussian military family. His military career was initiated by his father and cultivated by a superior military officer. He experienced relative deprivation in childhood, as well as an inferiority complex and depression as an adult.

Carl's father, Friedrich Gabriel, was a lieutenant in the Prussian army. He fought in the Seven Years' War, sustained a disabling wound, and was appointed a minor tax collector. On his meager income he had to support a family of eight (four sons and two daughters); the family "lived on the edge of poverty" (Paret 1976, 18).

Nonetheless, Friedrich Gabriel maintained many military ties and continued to regale Carl and his three older brothers with life in the Prussian army. "It was in this atmosphere of soldier's tales that Carl grew up" (Parkinson 1970, 22). In effect, by his own acknowledgment, Carl "grew up in the Prussian army" (quoted in Paret 1976, 17).

Carl joined the Prussian military at age twelve, his father having personally presented him to the commanding officer (Paret 1976, 19). "One reason for this early start to his military career was his father's lack of money, which made it difficult to keep him at home" (Parkinson 1970, 19). At about the same time, two of Carl's brothers joined the military, the fourth brother going to the seminary.

But military life was not to Carl's liking and he never found complete fulfillment. Outwardly, he was "assertive, positive, and entirely sure of himself," but inwardly he was "shy, nervous, and insecure" (Parkinson 1970, 18). "Clausewitz never fully belonged either to the dashing, duelling officer corps, or to the intellectual world of philosophy and scholarship. He tried to combine elements of both and was left an outsider. He had to be a part of the great Prussian officer society, where the first qualification was to be a nobleman [which at the time Carl was not]" (Parkinson 1970, 19). As a result, "Lauded by the Germans after his death, Clausewitz was sometimes scorned and often ignored by fellow Prussians during his life" (Parkinson 1970, 19).

(The "von," denoting nobility, went back to a great-grandfather, a title which had been abandoned long since. Reclaimed by Carl's father, it was not until 1827, when, in the light of the growing military stature of the three Clausewitz brothers, King Frederick William III formally recognized the family claim to noble status [Paret 1976, 15].)

In the meantime, Clausewitz felt isolated, rejected, and depressed. The more he studied, the more he became aware of his intellectual and educational inadequacy. He sought refuge in the Young Officers School in Berlin, whose entrance examinations he somehow managed to pass in 1801. "But his depression soon returned. Given his meager education, Clausewitz found difficulty following the lectures and keeping up with the other officer students. Once again he was an outsider. He could not afford to take part in all the student activities, even if he wished to do so. Clausewitz felt painfully misplaced and lonely. He was on the verge of leaving the army in despair" (Parkinson 1970, 32).

Clearly, Clausewitz needed someone older to rely on, a mentor. Then, in 1801, at the age of twenty-one, "he found a second father in [Gerhard von] Scharnhorst," who that year had been appointed a quartermaster general in the Prussian army (Paret 1976, 98). In time, Clausewitz maintained, Scharnhorst became "the father and mother and friend of my soul" (quoted in Parkinson 1970, 32). Scharnhorst guided the young officer through his military career and provided the necessary encouragement and opportunity. In fact, next to his own children, Scharnhorst came to consider Clausewitz as the closest person to him (Parkinson 1970, 32, 230).

Under Scharnhorst's tutorship, Clausewitz completed the Young Officers School. He also read the works of Kant, Voltaire, Montesquieu, and above all Machiavelli. Of the last, he said: "No reading is more important than the writings of Machiavelli" (quoted in Parkinson, 1970, 35).

On Scharnhorst's recommendation, Clausewitz was made aide-de-camp to Prince August of Prussia, thus entering Berlin society. But "Clausewitz could only be out of place in this society" (Parkinson 1970, 41). He was awkward, penniless, withdrawn, unable to relate or to adjust. Accordingly, as late as 1807, he acknowledged: "My life is a trackless existence" (quoted in Parkinson 1970, 19). "Twenty years later he was to believe he had been a failure" (Parkinson 1970, 19).

The Napoleonic wars of 1804–1815 ignited a burning ambition in Clausewitz, "an ambition for military glory worthy of Stendahl's Julien Sorel" (Howard 1983, 5). Given his intensely Francophobe sentiments, Clausewitz joined the Russian Army upon the Napoleonic invasion of 1812. A year later, he sought to rejoin the Prussian army but was rejected by King Frederick William III. "His rejection in 1813 affected him for the rest of his life" (Parkinson 1970, 214). For a time he was neither in the Russian army nor in the Prussian army—in other words, in a no-man's land. His depression came back. He joined the Russo-German Legion, which fought valiantly against Napoleon. In 1814 Frederick William III appointed Clausewitz an infantry colonel in the Prussian army.

In 1818, Clausewitz was promoted to major general and appointed director of the General War Academy. But his ideas for military reform were not accepted, leading to a recurrence of his depression. "By January 1823, Clausewitz was thoroughly depressed and disillusioned, and wished to give up his military career altogether" (Parkinson 1970, 304). Begun in 1810, he intensified his writing *On War* in the 1824–1830 period as a means of occupying his time and mind. But disillusionment intensified. He wrote in 1830 that, although by outward appearances he should be happy, "deep down in my soul I feel a profound melancholy" (quoted in Parkinson 1970, 325). He died on November 16, 1831. *On War* was published a year later.

Clausewitz's physicians diagnosed the cause of death as "a comparatively mild attack of cholera" (Parkinson 1970, 329). A Silesian newspaper reported the cause as "a nervous stroke" (quoted in Parkinson 1970, 330). A biographer believes he may have died of a heart attack (Paret 1976, 430). In any event, Clausewitz's widow Marie noted: "Life for him was a nearly uninterrupted succession of disappointment, of suffering, of mortification." He achieved much, she added, "But nevertheless he never reached the summit" (quoted in Parkinson 1970, 331).

HELMUTH VON MOLTKE, 1800–1891

Helmuth von Moltke was born to a Prussian military family. He experienced relative deprivation and love deprivation in childhood. Military life became a substitute for the home he lacked.

Helmuth was born in Parchim, Mecklenburg, on October 26, 1800. Shortly thereafter, the family settled in Holstein, then a part of Denmark.

Helmuth's father, Friedrich, was a lieutenant general in the Prussian Army, who subsequently transferred to the Danish army. A member of the impecunious aristocracy, he was in poor health, unstable, indifferent, and temperamental. As a result, he had great difficulty looking after a family of ten (six sons, two daughters).

The Napoleonic wars of 1804–1815 left a deep impression on Moltke and his generation, sowing the seeds of revenge.

When Helmuth was ten years old, his father Friedrich registered him and his younger brother Fritz in the Military Academy in Copenhagen, Denmark, where he had contacts and where the boys stayed for several years. In other words, a military career was forced upon Helmuth. In the meantime, the Danish experience produced culture shock and inferiority complex, leaving a lasting impression on the young boy (Kessel 1957, 12).

In 1814 Helmuth's mother Henriette left home and resettled elsewhere, effectively robbing the fourteen-year-old boy of any semblance of a home. Experiencing much hardship, living a "joyless youth" (Kessel 1957, 12), Helmuth became extremely inner directed and extremely ambitious. He became reserved and aloof; he developed an "iron diligence" and an "energetic will" (Kessel 1957, 13). The military academy became a substitute for home.

Under the influence of a Danish general, Moltke also developed an interest in things intellectual and spiritual. Homer and the Bible occupied the central spaces; he also read Shakespeare, Goethe, and Schiller (Kessel 1957, 11, 14, 15).

In 1820, Moltke joined the Danish army as a second lieutenant. Having been discharged, he returned to Berlin a year later and was commissioned in the Prussian army as a second lieutenant in 1822.

In 1823 Moltke reestablished contact with his mother Henriette, who had remarried. Once again, the mother became a source of emotional and financial support, enabling Moltke to concentrate on his military career (Kessel 1957, 29).

Later in the same year, Moltke entered the War Academy in Berlin. Ordinarily, an officer needed field experience for admission but, in view of Moltke's strong record, an exception was made.

Moltke was well received by the top echelon of the Prussian army, and much appreciated by his colleagues and superiors. As a result, he rose steadily to become chief of the Prussian general staff in 1857.

Moltke's victories in the Danish War of 1864, the Seven Weeks' War (with Austria) of 1866, and the Franco-Prussian War of 1870 sealed his fame. In 1871 he was promoted to the rank of field marshal.

PAUL VON HINDENBURG, 1847–1934

Paul von Hindenburg came from a long tradition of Junker military families, in which he always took pride. He was ashamed of his mother's middle class background, but could trace his paternal lineage to 1289 (Hindenburg 1920, ch. 1). That Hindenburg would become an army officer was charted for him merely by his status as the scion of an ancient (though impecunious) noble family.

A thoroughly unambitious man, Hindenburg rose to the highest ranks of the Prussian/German army, not by inner drive or tenacity but by circumstances and events. Thus Ludwig (1935) opens his massive biography of Hindenburg in the following fashion:

> My aim has been to show how an army officer was carried far beyond the limits of his potentialities, not by ambition, but by a "legend" which had accreted round his name. (Ludwig 1935, v)

Similarly, Wheeler-Bennett (1936) titled his biography of Hindenburg *Wooden Titan*, for the latter's lack of imagination and initiative, for frequently being dwarfed by persons and events, for his "self-protective coloring," for his rigidity, for his low intelligence, for his blind devotion to King and Fatherland. Hindenburg, Wheeler-Bennett writes, was "a giant among men, but a dumb giant" (Wheeler-Bennett 1936, x et passim).

Paul von Hindenburg was born on October 2, 1847, at Posen, a Prussian military fortress in a city of 185,000. His father served as an aide-de-camp to the fortress commander. Most of Paul's early childhood was spent going from one military outpost to another, as necessitated by his father's assignments, who eventually rose to become a major in the Prussian army.

In 1859, when the child was eleven years old, Paul's father personally delivered him to the Military Cadet Academy at Wahlstatt, Silesia. Hindenburg noted in his memoirs:

> That I should be a soldier was not the result of a special decision. It was a matter of course. Whenever I had had to choose a profession, in boys' games or even in thought, it had always been the military profession. The profession of arms in the service of King and Fatherland was an old tradition in our family. (Hindenburg 1920, 3)

And again: "It does not matter to what part of our German Fatherland my profession has called me; I have always felt myself an 'Old Prussian'" (Hindenburg 1920, 3).

A biographer strikes a more practical note: "The attraction of the Cadet Corps lay, then, not in the brilliant future it might dangle before a lad's eyes, but in the fact that the career of army officer afforded security for a lifetime" (Ludwig 1935, 30). Moreover, as we have seen, Hindenburg was always acutely conscious of the "honor" of serving the King and the Fatherland (Hindenburg 1920, passim; Ludwig 1935, passim).

ERICH VON LUDENDORFF, 1865–1937

As the child of a penniless, landowning family, Erich von Ludendorff's military career was preordained from the time he was born: "Ludendorff's life began in one era—Bismarck's—and ended in another—that of Hitler and the Third Reich. He was the child of the first and in many ways the unwilling father of the second" (Parkinson 1978, 11).

Family impoverishment left a deep impression on Erich, turning him into an austere figure, obsessive, compulsive, and without a sense of humor. Swept by the afterglow of the Franco-Prussian War, he determined from an early age to join the military as a means of serving the King and the Fatherland, and of improving his socioeconomic status.

As a child, Erich was "distinguished by only two unusual characteristics—his excessive cleanliness (he would not play with other children if he thought he might dirty his shoes)—and his passion for mathematics" (Goodspeed 1966, 12; cf. Parkinson 1978, 13). Erich had a "grim, thin-lipped mouth"; his "eyes were harsh, arrogant, staring . . . his rages flushed his face first to crimson and then to purple" (Parkinson 1978, 9). As a result, he remained aloof, isolated, and without friends.

Bent on a military career, Erich entered military cadet school in 1878, at age twelve. Five years later, he was admitted to the Military Academy at Lichterfelde. From 1893 to 1896 he was at the War Academy in Berlin. By 1914 he had been promoted to field marshal.

Ludendorff's entire military career was characterized by obsessive nationalism, compulsive behavior, and mania for work:

> Ludendorff's mind always seemed unable to accept compromise. He clamped himself to convictions and pursued them with an awesome doggedness. Later this trait would take fanatical form in his hatred of Jews and of German politicians whom he claimed stabbed Germany in the back by their acceptance of Allied terms in 1918–19. Now, as a teenager, he thought of nothing but work. Nothing must be wasted; no time must be lost, whether in the cold lecture halls, in his bare and inhospitable room, or on the parade ground and riding area. His every activity was directed toward being a soldier, able to fulfill the oath he had sworn on his sword-knot to serve the Kaiser and the Fatherland. (Parkinson 1978, 13)

Ludendorff's wife Margareth said of him: "Time was not reckoned in our house by hours, but by minutes" (quoted in Parkinson 1978, 19).

Ludendorff wrote in his memoirs:

> The homeland was not only the basis on which our proud military power rested, and which must therefore be carefully safeguarded; it was the life-giving source which had to be kept clear, pure and yet potent, lest it lose anything of

that virtue wherein it steeled the nerves and renewed the strength of the Army and Navy. (Ludendorff 1919, I:3)

And again: "I am neither a 'Reactionary' nor a 'Democrat.' All I stand for is the prosperity, the cultural progress and national strength of the German people, authority and order. These are the pillars on which the future of our country rests" (Ludendorff 1919, I:8). And again: "Mine has been a life of work for our country, the Emperor and the Army. During the four years of war I lived only for the war" (Ludendorff 1919, I:13).

In 1914 Ludendorff was appointed chief of staff to Paul von Hindenburg. He observed:

I was proud of my new task and of the trust placed in me.... I was exalted at the thought of serving my Emperor, Army and Fatherland, in a position of great responsibility at a most critical point. Love of country, loyalty to my Sovereign, appreciation of the truth that the duty of everyone is to devote his life to his family and the State, this was the heritage which I took with me from my home and my portion in life. (Ludendorff 1919, I:42)

Following retirement from the German army, Ludendorff exhibited Nazi sympathies, was admired by Hitler, and was elected a Nazi deputy in 1925. Upon his death on December 20, 1937, a funeral procession moved through Munich: "Now Hitler walked behind the coffin, and Field-Marshal von Blomberg, the Nazi War Minister, delivered the oration, and a huge swastika lay over the casket guarded by four Nazi soldiers" (Parkinson 1978, 229).

ERWIN ROMMEL, 1891–1944

Given the afterglow of the Franco-Prussian War, given the creation of the German empire, given Germany's quest for global supremacy, given the constant glorification of the Fatherland, it is not surprising that "Even while he was at school [in his teens] Rommel's dream was to be a soldier" (Mellenthin 1977, 55). In other words, Rommel's military career was entirely situational.

Erwin's father, also named Erwin, who had served briefly as an artillery lieutenant, was a mathematician/schoolmaster, as was his father before him. Not being intellectually oriented, unable to follow in his father's footsteps, seeking social and financial security, the young Rommel decided to join the military. He realized at the same time that, not being a member of the Prussian aristocracy, promotions would be slow in coming. Accordingly, his ascendancy to the top both delighted and stunned him.

As a child, Erwin was puny and inactive, given to introspection and daydreaming. "Then, as a teenager, he suddenly shook off his lethargy. He started a program of exercise and systematically developed his body. . . . Rommel played tennis, bicycled, skied, and learned how to skate. He acquired many of the characteristics common among Wuerttembergers— toughness, self-reliance, stubbornness, pragmatism, and thrift" (Mitcham 1984, 28; cf. Young 1950, 12–13).

Having determined to become a soldier, the young Rommel faced little or no opposition from his family. Erwin's application to the engineer branch of the German army was rejected, as was his attempt to join the artillery. At last he was accepted by the infantry, which he joined in July 1910. "Finally, in March 1911, he achieved one of the major ambitions of his life: He was sent to the War Academy at Danzig to undergo officer training" (Mitcham 1984, 29; cf. Irving 1977, 10; Young 1950, 29–30).

Rommel did rather well at the academy. His commandant characterized him as "firm in character, with immense willpower and a keen enthusiasm. . . . Orderly, punctual, conscientious and comradely. Mentally well endowed, a strict sense of duty" (quoted in Mitcham 1984, 29).

In 1917, following a series of successful campaigns, Rommel was "decorated with the *Pour le Mérite,* a medal roughly equivalent to the American Congressional Medal of Honor. . . . He wore this medal [which came to be known as the 'Blue Max'] with the greatest pride until the day he died" (Mitcham 1984, 35).

In general, the battlefield transformed the young Rommel into a warrior of the first order. He became the quintessential fighting machine: decisive, swift, daring. "From the moment that he first came under fire he stood out as the perfect fighting animal: cold, cunning, ruthless, untiring, quick of decision, incredibly brave" (Young 1950, 16). "He was the body and soul of war," one of his superiors commented later (quoted in Mitcham 1984, 30). And so emerged the legend of the "Desert Fox."

In 1942, when Hitler promoted Rommel to the rank of field marshal, Rommel is reported to have written his wife Lucie-Maria, telling her that "he would rather the Füehrer had presented him with a single fresh division" (Mellenthin 1977, 76; cf. Young 1950, 110). However, he also wrote his wife that "To have become a field marshal is like a dream to me" (quoted in Irving 1977, 186). A biographer comments: "To become a field marshal was to become an immortal" (Irving 1977, 186).

Though a conservative nationalist and for a time a supporter of Hitler, Rommel never became a member of the Nazi party. Rather, he became increasingly critical of Hitler as Nazi atrocities mounted. He was suspected of the July 20, 1944 complicity to assassinate Hitler. On Hitler's orders, he was given the option of either taking poison or facing trial by the "People's Court." Rommel took the former course, ending his life on October 14, 1944, at age fifty-three.

Chapter 8

Other Leaders

MOSHE DAYAN, 1915–1981

Given Moshe Dayan's life span, given the aftermath of the two world wars, given the turbulence of the times in Palestine, and given the Jewish quest for a homeland, every Jewish person was involved in intense nationalism (or Zionism), and every Jewish life had a military component. Dayan acknowledges: "The first few years of my childhood were the war years, and they were grim for all of us" (Dayan 1976, 28).

Moshe's parents—Shmuel and Dvorah—had emigrated from two nearby villages in the Ukraine in 1908 and 1913, respectively. They met at Degania, a communal farming settlement in Palestine. They married in 1914, upon learning that Dvorah was pregnant (with Moshe) (Teveth 1972, 1–2).

(We are not informed of Shmuel's Ukranian surname. We are told, however, that he adopted the name "Dayan," meaning a rabbinical judge in Hebrew, upon reaching Palestine [Dayan 1976, 23; Teveth 1972, 5].)

While Shmuel was a thoroughly committed Zionist, Dvorah wavered. From her youth she was given to leftist causes. Cosmopolitan by nature, she was "bored" with life in Degania. Rather than working to advance the cause of Zionism, she preferred to study Russian literature, particularly Chekhov and Tolstoi. She wanted "to see the world." She spoke Hebrew haltingly. As a result, she was not fully accepted by the Degania community; she was considered an outsider and an "alien" (Teveth 1972, 1–2, 9–13).

Moshe grew up in a rather stormy family environment. An ambitious Zionist, Shmuel, a farmer-turned-activist, was frequently away on political and organizational work throughout the Middle East, Europe, and the United States. Dayan noted that as early as 1921 (when he was five years old):

> My father had always had an urge to play a part in public life. He became active in party and organization work and was sent on missions abroad—at least twice for nearly a year at a time. During these absences, my mother had to carry the full burden of the farm work, with what assistance I could give her. Looking back I realize more clearly than I did at the time with what fortitude and tenacity she carried on, in spite of poor health, chronic debts, and a growing family. (Dayan 1976, 30)

Tension developed between Shmuel and Dvorah and for a time there was talk of divorce (Teveth 1972, 41–42, 73). At one point Shmuel wrote Dvorah: "I feel that I cause you nothing but sorrow and that you haven't known a single moment of happiness with me" (quoted in Teveth 1972, 23).

These family circumstances produced several consequences for Moshe Dayan. Above all, he developed an ambivalent attitude toward his father, while growing ever closer to his mother. On the one hand, he admired—and closely followed—the activist model established by his father: "Thus from his earliest childhood, Moshe was raised on the teachings of Zionism and its historical justification" (Teveth 1972, 33; cf. Dayan 1976, 22–23 et passim). On the other hand, "Moshe felt his father's absence keenly and was even more sensitive to his mother's hardships" (Teveth 1972, 40). Indeed, Moshe became increasingly disappointed in his father and "deeply impressed by . . . [his mother's] decisiveness and daring" (Teveth 1972, 61). As early as age fifteen (1930), he recorded in his diary:

> Even as a child I understood that [father] talked in formulas. . . . When a son discovers his father talking nonsense, he begins to disrespect him. There was no point in arguing with him because he never developed his thoughts logically. Arguments with him were not of the kind in which one side or the other could be convinced. Father would trot out truisms, and that was that. With mother, one could discuss things and argue. She could prove to you that you were wrong or admit that you were right. (quoted in Teveth 1972, 61)

Moreover, Dayan became anxious to leave an unhappy household, marrying his first wife Ruth Shwarz in 1935, at age twenty:

> Yuka [an old girlfriend of Moshe's], as well as some of Moshe's friends, believed that one of the main motives for his wish to marry so young was his desire to leave his parents' house. Israel Gefin, who later married Aviva [Moshe's sister] described the Dayan household as oppressive. Shmuel preached inces-

santly and Dvorah always walked about the house despondently, "the very embodiment of suffering on the face of the earth." (Teveth 1972, 77)

Beyond this, family circumstances intensified Moshe's inherited activism and propelled him into the political-military arena. As early as 1929, Moshe joined the Haganah, an underground paramilitary organization; at fourteen, he was the youngest member of the group. In 1939 the British declared the Haganah illegal and arrested Dayan and forty-two of his comrades. Sentenced to five years in prison, Dayan served only one and a half years, rejoining the British and Zionist forces for military activity. Following a series of successful military maneuvers, he was promoted to major general in 1949. In 1952 he attended senior officers school at Devizes in England. A year later, he was all set to assume the position of chief of staff.

Moshe Dayan is one of few military leaders to have had little formal military training, learning everything first-hand in the course of the struggle. He did have a couple of mentors and role models: Yitzhak Sadeh (an underground resistance leader) and Captain Orde Charles Wingate (the British guerrilla warfare instructor). Beyond this, Dayan was a self-made man.

Finally, given Dayan's life circumstances, it is not surprising that he developed a very high opinion of himself. Even in his teens, his classmates noted "the Dayan trait of haughtiness, contempt for others ... his sense of superiority. This haughtiness was a Dayan trait" (Teveth 1972, 53). Dayan himself noted: "Emotional partnership, sociability, and absolute egalitarianism were not in keeping with my nature" (Dayan 1976, 40).

GIUSEPPE GARIBALDI, 1807–1882

Giuseppe Garibaldi was the nineteenth-century Italian counterpart of Moshe Dayan, just as Moshe Dayan was the twentieth-century Israeli counterpart of Giuseppe Garibaldi. Both were motivated by the elemental considerations of nation-building and national consolidation. Like Dayan, Garibaldi had virtually no formal military training—he polished his military skills on the battlefield. Unlike Dayan, Garibaldi was entirely free of haughtiness and vanity.

Garibaldi was born in Nice, France, in 1807, to a family of fishermen and coastal traders. In his youth he was a sailor for about ten years (ca. 1822–1832) in the Mediterranean and the Black Sea. In 1832 he obtained a master's certificate as a merchant captain. From 1833 onward, he became an integral part of the Italian Risorgimento.

The Risorgimento dates from the disintegration of Italy following the Napoleonic invasion of 1796–1797 to the reunification of Italy in 1870.

The triumvirate of the Risorgimento were: (1) Giuseppe Mazzini, the intellectual leader; (2) Count Camillo di Cavour, the diplomat/political leader; and (3) Giuseppe Garibaldi, the military leader. When Garibaldi met Mazzini in 1833, he immediately joined the latter's underground nationalist organization, Young Italy.

Soldier and sailor, Garibaldi's sole and unwavering objective was the unification of Italy—followed, ideally, by a federation of European states. He traveled throughout Europe, the United States, and Latin America, everywhere preaching the gospel of Italian unification. Years of exile and imprisonment never deterred him or shook his faith (Hibbert 1966, passim; Larg 1970, passim; Smith 1969, passim; Viotti 1979, passim).

In 1855, Garibaldi presented his program for Italy in the following terms:

> To make *a single Italy* must be our first goal. The Italian peninsula is made up of small states: there is Tuscany as well as Piedmont; there are some Italians who owe loyalty to the Pope, others who acknowledge the Bourbons [of Naples], others who are republicans, and others who look to Murat [French Pretender to Southern Italy]. Besides these there are some other even smaller groups who, however negligible, cannot help but damage the concept of national unity. All these elements must amalgamate and join whoever is strongest among them, or else they will be destroyed; there is no middle way. The most substantial element in Italy I take to be the Piedmontese, and I therefore advise that all should gather round them. We should be ready to accept a rigorous dictatorship from Piedmont as a means of emancipating ourselves from foreign domination. (quoted in Smith 1969, 29–30; italics and bracketed matters in original)

In 1859, upon returning to Piedmont following a tour of Europe, Garibaldi restated his objectives:

> You have just narrated my history, and it is now my part to tell you how proud and happy I am to find myself again among this brave people, of whose courage and attachment I have experienced so many proofs. I repeat to you that, to the last moment of my existence, I shall be devoted to my country, body and soul. For fourteen years I have served the cause of liberty in foreign lands without pay or reward. What then will I not do for my native country? Events are progressing favorably, but there is still much to be done. The day is come when Italy shall regain her complete independence. This time it must be accomplished, and from the Alps to Sicily she must be free. Providence has given us the man we needed to re-knit us together. It is round Victor Emmanuel that we must rally to repulse the stranger from our soil. We will no longer bear the foreign yoke. Let but our oppressor retire and leave us to enjoy our possessions in peace, and we will at once welcome him as a friend; but so long as he desires to subject us to his dominion he has nothing to expect from us but the fire of our artillery. It is only by union and strength that we shall obtain our freedom. (quoted in Larg 1970, 220)

Giuseppe Garibaldi was a man entirely devoid of personal ambition. He had no interest in worldly recognition, high office, or fortune. When he formally turned over some conquered territories to King Victor Emmanuel on October 26, 1860, he refused the rank of major general, the title of Prince of Calatafimi, a large pension, a castle, and nomination of his son, Menotti, as the king's aide-de-camp. All he took with him was a sack of seed, some coffee and sugar, a bale of fish, and a supply of macaroni (Viotti 1979, 129).

Decades of hard work by Garibaldi, Mazzini, and Cavour finally paid off. Italy was reunified under King Victor Emmanuel of Piedmont in 1870.

ISOROKU YAMAMOTO (NÉ TAKANO), 1884–1943

Isoroku Takano was born to a family of samurai in a city of seafaring men. He experienced relative deprivation and identity crisis in childhood and youth. When opportunity presented itself, he adopted the family name Yamamoto as a means of enhancing his social status and military career. His military experiences intensified his nationalism.

Isoroku was born in Nagaoka, a medium-sized city on the northern shore of Honshu Island; he was the seventh and youngest child of Sadayoshi Takano, a former samurai. "For hundreds of years, Nagaoka was the home of seafaring men. The Nagaoka clan produced strong fighting men, or samurai, and one family of that clan, the Takano family, was particularly notable for the strength and character of its samurai" (Hoyt 1990, 16).

Following the civil war of 1868—the Bosshin War—the Takano family, including Sadayoshi, fell on hard times, earning a meager living and wandering on Honshu Island. Isoroku's childhood and youth were spent in hard work in order to help maintain the family, go to school, and find the opportunity to escape his life of poverty and hardship. In middle school he developed a particular interest in gymnastics, which he practiced regularly. Among other things, his teacher staged the Bosshin War on the athletic field, allowing Isoroku and his classmates to live that war all over again (Hoyt 1990, 19).

In time Isoroku became highly sensitive to his family's poverty and his position as the youngest son:

> Isoroku was, however, suffering from what would later be termed an "identity crisis," created by Japan's familial system, in which the eldest son is heir apparent to the position of head of family. All four of Sadayoshi Takano's children by his first wife were sons, and, of course, they were grown men by the time their father married again and the second batch of three children began coming along. . . .

As a small boy, Isoroku had not given a lot of thought to such matters as inheritance and family status, but the matter was pushed rudely to his attention when he was fourteen years old. That year his oldest half-brother, Yuzuru, died. The whole family attended the funeral, and Sadayoshi took that occasion to tell Isoroku that his position as youngest son in the family meant that he could never expect anything at all. (Hoyt 1990, 23)

This came as "a terrible shock" (Hoyt 1990, 23).

Having been born in a seafaring environment, and lacking family money and position, a naval career seemed a natural course for Takano to follow. Accordingly, having scored second on the competitive entrance examinations, he entered the Imperial Naval Academy in 1901, graduating four years later. He fought with distinction in the Russo-Japanese War of 1904–1905, an experience which intensified his growing nationalism and loyalty to the emperor.

The deaths of both parents in 1912 caused Takano much distress, but he managed to overcome it. He compensated in part by attending the Naval Staff College from 1913 to 1916.

Upon graduation, Isoroku Takano faced the unique opportunity of becoming the leader of the Yamamoto family, "an honorable and ancient one in the history of Japan" (Hoyt 1990, 37). The last leader of the Yamamoto clan had been killed in the Bosshin War, jeopardizing the very future of the family, since he had no sons.

For several years the family had looked for a new leader, and Isoroku's performance had been so impressive that they asked him to take the family name and the responsibility. Mother and father dead, his elder brothers standing between him and leadership, there was no conceivable reason for Isoroku to refuse this considerable honor and the emoluments it brought him from an important clan. So in 1916 he became Lieutenant Commander Isoroku Yamamoto of His Majesty's Imperial Japanese Navy, and the change was duly recorded in the Yamamoto clan records. (Hoyt 1990, 37; cf. Agawa 1979, 67)

As Yamamoto's career advanced, culminating in the rank of admiral in 1940, his nationalism grew increasingly fierce and his commitment to the emperor and the country became unqualified. To die for the emperor and the country, he maintained, guarantees immortality (see Hoyt 1990, 102).

For his role in planning the Pearl Harbor attack, Yamamoto was ambushed and killed by American fighter planes on April 18, 1943. The order for his assassination came directly from President Franklin D. Roosevelt (Hoyt 1990, 248).

HIDEKI TOJO, 1884–1948

Hideki Tojo's military career was charted for him by a strong military family tradition, which included fierce nationalism.

OTHER LEADERS 103

Hideki was born to a family of samurai. His father, Hidenori, had been a general in the Japanese Imperial Army. Hideki had no choice but to follow in his father's footsteps:

If there had ever been any possibility that Hideki Tojo would have been anything other than a soldier of Japan, it was eliminated very early in his life. He was born on the 30th of December, 1884, into the family of Hidenori Tojo, a dedicated career soldier who had joined the Japanese Imperial Army at the age of sixteen. By the childhood deaths of two elder brothers, Hideki became the eldest surviving son. Thereafter, tradition and all the disciplines of Japanese family structure required him to be a soldier as his father was, carrying on a profession which, in Japan, had long enjoyed a unique and privileged status. (Browne 1967, 7–8; cf. Butow 1961, 7, 26)

Hideki entered the Imperial Military Academy in 1902, graduating four years later. Having completed Staff College in 1915, he continued a successful military career that made him a general by 1941.

Even as a young man, Tojo was obstinate, opinionated, intemperate, and combative (Browne 1967, 20). These attributes intensified as he matured. He was utterly absorbed in his profession. "Hobbies or sports he had none and, he always insisted, his hobby was his work" (Browne 1967, 30).

Like Isoroku Yamamoto, Hideki Tojo was entirely dedicated to nationalism and patriotism: emperor and country came first and foremost. In addition, there was an element of racism:

Of all the oriental peoples, the Japanese are the most racially conscious. Indeed, the notion of the ethnological superiority of the Japanese as pure in strain and divinely descended was centuries old in Japan long before the first nineteenth century European exponents of the myth of Aryan supremacy were born. (Browne 1967, 31)

Tojo was subjected to "ultranationalism" from his earliest days:

Indoctrination in ultranationalist principles became part of all school curricula. All the myths of Japan from its miraculous creation from the Sun Goddess two thousand years before were taught as historical fact. *Hakko ichiu*—the injunction of the god-ancestors that the eight corners of the world must be united by Japan "under one roof"—was inculcated as an article of faith. Pupils, from toddlers armed with miniature rifles, were trained by regular or retired noncommissioned officers of the army and navy. (Browne 1967, 110)

It is not surprising that Tojo subscribed to the concept of *kokutai*, "a mystic and untranslatable term signifying the oneness of the Japanese state and people with their divinely descended Imperial line" (Browne 1967, 30).

These complementary elements of nationalism, mysticism, and emperor worship came to a head during World War II. Tojo blamed the Allies for everything: Britain and America had provoked the war; they had imposed economic sanctions on Japan; they had provided aid to China; they had prevented Japan from saving China from communism; and so on (Browne 1967, 228).

Following Japan's surrender in 1945, Tojo, then prime minister, was arrested as a war criminal. Following an unsuccessful suicide attempt, he was tried before a military tribunal, found guilty, and hanged on December 23, 1948.

ANTONIO LOPÉZ DE SANTA ANNA, 1794–1876

(A devout Catholic, Antonio used Santa Anna—rather than Lopéz—as his surname because Santa Anna indicates a devotion to Saint Anne [Callcott 1964, 4; Jones 1968, 21].)

Santa Anna was a child of his chaotic environment. The dual forces motivating him were a desire for personal glory (egomania) and for the glory of Mexico.

Santa Anna's life reflects Mexico's turbulent history in the nineteenth century, which represented an "era of chaos, revolution, changing alliances, opportunism, diplomatic intrigue, and almost continual warfare" (Jones 1968, 20). Accordingly, Santa Anna "developed an early interest in military life" (Jones 1968, 21). He wrote in his autobiography: "I have, since my earliest years, been drawn to the glorious career of arms, feeling it to be my true vocation and calling" (Santa Anna 1967, 7).

Santa Anna had very little formal education and no formal military training whatever. As did some other leaders, he mastered military science on the battlefield. Having joined the army in 1810, he was promoted to the rank of general in 1829, at the tender age of thirty-five.

On the one hand, Santa Anna was a colorful, dashing, and charming military figure. On the other, he is variously described as ambitious, egotistical, despotic, unprincipled, and cruel—addicted to opium and cockfighting (Callcott 1964, passim; Hanighen 1934, passim; Jones 1968, passim). Above all, he saw himself as "The Napoleon of the West" (this description constitutes the subtitle of Hanighen 1934). He took Napoleon "as a model; he even arranged his hair from back to front as the Little Corporal had worn his when crossing the Alps, and he bought a white charger resembling the one his hero had always used" (Jones 1968, 24).

Vanity and nationalism alternated, and they were the dual motives that animated Santa Anna. He is utterly candid in his autobiography:

OTHER LEADERS 105

> In the hearts of most men there lurks a sentiment which they carefully try to hide from their fellow men. This foolish sentiment is that which causes man to aspire constantly to immortality. Not all men, however, succeed in inscribing their names on the walls of the temple of glory.
> I cannot deny that when I was young the idea of glorifying my name passed through my mind. I aspired to perform noble deeds that would live forever in the hearts of men. In later years, however, a nobler sentiment, lacking all personal ambition, took its place and constantly directed all my actions. This sentiment was the love of liberty and the desire to glorify the name of my country. To this love of country which I have avowed, and to nothing else, I owe my transitory rise to power.
> I desired to spend my life performing noble deeds for that magnificent country where I was born and reared. But the constant upheavals and surges of revolution opposed me. And, in their constant drive for power, not all the men who surrounded me could measure up to those aspirations of patriotism which motivated me. (Santa Anna 1967, 3–4)

And again: "I ... occupied myself solely with helping Mexico become a nation. ... I have dedicated myself entirely to the noble profession of defending my country" (Santa Anna 1967, 5).

Santa Anna commented on the abdication of Emperor Augustin I (with whom he had clashed) in 1823: "My victory could not possibly have been more splendid! Judge and jury that I was in those momentous times of the destiny of my country, I remained faithful to every promise in the program that I had proclaimed for the Republic. With a zeal that was almost religious in nature, I followed it to the letter!" (Santa Anna 1967, 17–18).

Following the abdication, Santa Anna was appointed provisional governor:

> Cannons hailed my arrival into the port, and the jubilant people demonstrated in my honor. The military commandant, Lieutenant N. Roca, quickly placed himself at my command. The opposing colonel, Benito Aznar, who had been steadily besieging the city, quickly followed his example. Both the people of Campeche and Merida overwhelmed me with honors, and the Provisional Junta elected me political governor of the entire province. By settling the differences between the cities in peace, I successfully established order and restored security to the region. I organized active committees and permanent government bodies. In addition, I built up the fortifications, and, in every manner, I provided for the security of the province. (Santa Anna 1967, 18–19)

Santa Anna had lost a leg to a cannon in the 1838 war with France. His self-glorification reached an apotheosis when he had the leg exhumed and reburied with great ceremony five years later:

It is perhaps not surprising, then, to find a most singular ceremony taking place on the 26th of September, 1843. In the cemetery of Santa Paula a large crowd had gathered in spite of a broiling sun and between two files of grenadiers a procession of officials and military bore a small urn of crystal. It contained Santa Anna's leg dug up from its resting place in Manga de Clavo [Santa Anna's family home] and brought hither for a formal interment. A gilded column ornamented with the inevitable inscriptions and fasces awaited it and after placing the urn on the capital, a small stone cannon topped by a Mexican eagle was superimposed. (Hanighen 1934, 179–80)

GRIGORY ALEXANDROVICH POTEMKIN, 1739–1791

Grigory Potemkin was born to a military family of lower aristocracy. In some ways he had a stormy childhood. Having resolved to pursue a military career—and having established a liaison with Catherine the Great—his words and actions became increasingly governed by national and personal ambition.

Grigory's father, Alexander, a member of the lower nobility, had risen to the rank of colonel in the Russian army. But family life was far from smooth: "During his father's lifetime Grigory had a difficult childhood." There were "constant disputes" between the parents, and the father was too poor adequately to provide for his six children (Soloveytchik 1947, 42).

The father died in 1746. Grigory, then seven years old, was placed in the care of a cousin who was also his godfather.

In his childhood, Potemkin was torn between a career in the military and one in the ministry—but always in a leadership role. While in grade school, he told his classmates: "If I become a general, I shall have soldiers under my orders; if I become a bishop, it will be monks" (quoted in Soloveytchik 1947, 44). By his mid-teens Grigory had resolved this problem: as befitting a nobleman in eighteenth-century Russia, he joined the military in 1755.

By 1759, Potemkin had been promoted to the rank of captain:

In St. Petersburg he was able to lead the life that was customary for young noblemen serving the army. Drinking, gambling, and promiscuous love-making took up most of their time, and Potemkin appears to have plunged into this dissolute life with special and boundless zest. Being a poor man, however, and not wishing to be in any way behind his more prosperous friends, he soon piled up very considerable debts. But he moved "in the best of circles," enjoying great popularity. (Soloveytchik 1947, 45)

Then in July 1762 a transformation took place. Potemkin participated in a coup that toppled Czar Peter III and helped Catherine II to ascend the throne. As Potemkin became increasingly close to Catherine, nationalism and ambition took over. Potemkin's overriding objectives were to assure the grandeur of Russia, to serve Empress Catherine, and to give play to his personal ambition. Specifically, as he put it, he developed a "particular zeal towards the person of Your Majesty" (quoted in Soloveytchik 1947, 54).

In 1774, Potemkin and Catherine began their legendary liaison, further fanning Grigory's nationalism and ambition:

> From the very outset Potemkin could see that his passion for this woman, which had now lasted for a good many years, was not only reciprocated completely, but that Catherine's love for him was as stormy in character as his own feeling for her. . . . To work for her was to work for Russia, her glory was Russia's glory. To dominate Catherine meant to dominate the whole Russian Empire. (Soloveytchik 1947, 78–79)

Soon, Catherine was consulting Potemkin "about everything, from the state affairs of the highest importance to the most trivial court and personal matters" (Soloveytchik 1947, 131). As for Potemkin, "He was, as ever, unable to bear the thought of Catherine taking any decision, whether important or trivial, without first consulting him, and she, for her part, was only too glad to go on sharing the vicissitudes of power and responsibility with a consort who had proved such an able adviser and collaborator" (Soloveytchik 1947, 177).

Potemkin continued to consolidate his power and position until he became the de facto ruler of Russia during the 1770s and the 1780s. Catherine honored Potemkin by naming him the Prince of Tauvis, the ancient name of Crimea, whose annexation Potemkin had engineered in 1783.

ANTON IVANOVICH DENIKIN, 1872–1947

Anton Denikin was born to an impoverished military family. He experienced relative deprivation in childhood. Following the example of his father, he became entirely dedicated to Russia, the czar, and the military.

Denikin's father, Ivan, a serf, was handed over by the estate owner as a military recruit at age twenty-seven in 1834. He rose in the Russian army to the rank of major, retiring in 1869, three years before Anton was born.

The father, whom the son describes as "an extremely zealous soldier" (Denikin 1975, 5), brought home stories of military life to which Anton would "listen with rapt attention" from his earliest days (Denikin 1975, 5). When the father volunteered for the Russo-Turkish War of 1877–1878, mother and child cried: "But in the depth of my little soul I was proud that my papa was going to the war" (Denikin 1975, 7).

Years later Denikin wrote in his memoirs:

> During my first year, according to the ancient custom, my parents practiced divination on some family holiday or other. They placed on a tray a cross, a toy saber, a wineglass, and a little book. Whatever I first touched was supposed to predestine my career. When they brought them to me, I first was drawn to the saber....
>
> The ... saber, as a matter of fact, actually predicted my life's career....
>
> Father's tales and my childish play ("war" with saber and rifle) all built toward an agreeable decision. As a boy I used to disappear for hours to the exercise area of the 1st Rifle Battalion and walk to the water troughs and horse baths with the Lithuanian lancers. They allowed me to shoot in target practice with the frontier guards.... The bullets whistling over our heads were terrifying but always interesting and awe-inspiring to a small boy, eliciting envy for the soldier's life. Walking back along the road with the riflemen, I accompanied them in soldiers' songs. (Denikin 1975, 30–31)

Anton lived in poverty in childhood and youth. During the first years of his life, the family (father, mother, grandfather, nursemaid, and Denikin) lived on Ivan's pension of thirty-six rubles per month. When Ivan died in 1885 (Anton was thirteen), the family's financial situation took a sharp turn for the worse: "With Father gone our material circumstances seemed catastrophic. Mother began to receive a pension of only twenty rubles a month" (Denikin 1975, 29). Anton recalled a specific incident when he was in the fifth grade: "And although I still remember how good the steamed 'little hearties' (kolbasy sausages) smelled as I stood near the buffet counter in the school corridor during midday break, they were beyond my means" (Denikin 1975, 9).

Hardship notwithstanding, Anton's devotion to Russia, the czar, and the army never wavered. The translator/annotator of his memoirs described Denikin as "a man with an exalted concept of duty and a devotion to the military calling that amounted to a sense of mission ... completely devoted to his ... country.... Profoundly influenced throughout his life by his father's devotion to Russia ... Denikin never wavered from his childhood determination to become an army officer" (Margaret Patoski, "Introduction," Denikin 1975, vii–vii). Accordingly, having completed *realschule* (a type of combined primary/secondary education), Denikin attended the Kiev Junker Military School (1890–1892) and the General Staff Academy at St. Petersburg (1899–1901). He rose rapidly to become a lieutenant general by 1915 and the commander of White Russia (the "Volunteer Army") three years later.

At the beginning of August, 1918, rumors of the slaying of the imperial family in Ekaterinburg reached the Volunteer Army, and they were soon proved to be correct. The impact of the news was overwhelming. Despite the systematic terror engaged in by the Soviet regime, despite the cruelties of the civil war, which had blunted the sensibility of many, this murder brought home with extraordinary force all the savagery of the lawlessness and arbitrary rule which was sweeping the country, with no mercy left for women and children.

General Denikin ordered memorial services to be held. Officers and men of the Volunteer Army prayed with fervent concentration for the souls of the imperial family. (Lehovich 1974, 245–46)

With the cause lost, Denikin emigrated to England and the United States. He died in Ann Arbor, Michigan, in August 1947.

GEORGI KONSTANTINOVICH ZHUKOV, 1896–1974

Born to a family of poor peasants, Georgi Zhukov experienced serious relative deprivation in childhood. As a result, he came to view the military as an avenue of social advancement. Not surprisingly, moreover, as he grew up Zhukov found his sympathies to be with the Bolsheviks. Accordingly, he set out to glorify the Soviet state and to seek personal glory as well.

Georgi lived a life of utter poverty and hardship in childhood and youth. In his memoirs he refers to "our desperate financial circumstances.... [W]e, the children of the poor, suffered together with our parents" (Zhukov 1971, 12). Our house "looked the worst in the village" (Zhukov 1971, 12).

Between 1905 and 1914 (ages nine to eighteen), Zhukov was sent to Moscow to apprentice as a furrier and a garment maker. In 1915 he was drafted into the Russian army as a private. In 1919, immediately following World War I, he joined the Communist Party. A year later he became a member of the Razan Cavalry Corps. In 1924–1925 he attended the Higher Cavalry School in Leningrad. In 1929–1930 he was sent to the Frunze Academy Cavalry Inspectorate in Moscow. Following a series of rapid promotions, he was named a marshal of the Soviet Union in 1943.

As Zhukov matured, he expressed increasingly anti-czarist and pro-communist sentiments. He wrote in 1916:

Under the yoke of the Czar, whose mad recklessness had brought on these three years of bloodshed, the peasants had reached the end of their tether. The soldiers already understood that they were being sacrificed in the interest of the powers-that-be, in the interests of those who had always fleeced and robbed them right and left. (Zhukov 1971, 41)

By contrast:

The Bolsheviks, as we heard, were struggling against the Czar to secure peace, liberty, and happiness for all the toiling folk. By now the soldiers themselves were pressing for the termination of the war.

Though I was an NCO, the privates trusted me and we often discussed serious topics. Of course I was still politically naive, but I realized that only the Bolsheviks could give the Russian people peace, land, and liberty. I tried to impress this on my subordinates. And they repaid me for this. (Zhukov 1971, 42)

Zhukov reiterated these sentiments at every possible opportunity. Throughout the 1930s and the 1940s, he waxed eloquently about the accomplishments of the Soviet state in the areas of economy, industry, agriculture, art, culture, literature, and the like (Zhukov 1971, passim). At the end of World War II, Zhukov went so far as to claim that the Soviet Union had single-handedly defeated the Nazis (Zhukov 1971, 107).

Zhukov's glorification of the Soviet Union went hand in hand with a program of self-glorification and egomania (Zhukov 1971, passim; Chaney 1971, passim). In this, he found constant encouragement from party functionaries who urged him to seek ever higher office. This quest culminated in a meeting with none other than Stalin himself in 1941:

After a word of greeting, Stalin said:
"The Politbureau has decided to relieve Meretskov from the duties of the Chief of the General Staff and to appoint you in his place."
I had been expecting anything but this sort of decision and in my confusion I said nothing. Then I replied:
"I have never worked in staff before. I have always been in the field. I cannot be Chief of the General Staff."
"The Politbureau has decided to appoint you," said Stalin, laying special emphasis on the word "decided."
Seeing that objections were pointless I offered my thanks for the trust placed in me and added:
"But if I don't make a good Chief of Staff, I shall ask to be transferred back into the field." (Zhukov 1971, 187)

In his quest for glory, Zhukov eventually clashed with Stalin and the Communist Party: he had become too popular for Stalin's taste. Accordingly, in 1946 Zhukov was dismissed from the Central Committee. A year later, the Central Committee officially stripped him of all his political and military posts, issuing a statement reading in part:

Zhukov imagined that he was the sole hero of all the [World War II] victories achieved by our people and their armed forces under the Communist Party's leadership, and he began flagrantly violating the Leninist Party principles of leadership of the armed forces.... He proved to be a politically unsound person, inclining toward adventurism both in his understanding of the primary objective of the Soviet Union's foreign policy and in his leadership of the Ministry of Defense. (quoted in Chaney 1971, xi)

Zhukov was partially rehabilitated during the Khrushchev era and reappointed Minister of Defense in 1957. But the rehabilitation did not withstand the tumult of post-Stalinist politics. Zhukov died in 1974.

FRANCISCO FRANCO, 1892–1975

Francisco Franco was born to a military family. Abandoned by his father, he grew closer to his mother while at the same time rapidly steering toward a military career. Military training, in turn, nurtured and fanned his budding ultranationalism.

Francisco was born and raised in El Ferrol, a navy town in Galicia in northwestern Spain, the son of Nicólas Franco and Pilar Bahamonde. His paternal family had had a tradition of naval service going back to the early eighteenth century; his mother also came from a naval family (Lloyd 1969, 19; Crozier 1967, 29). "Don Nicólas Franco . . . became the fourth in the direct male line to become . . . [a naval] officer . . . when he was commissioned in 1878" (Trythall 1970, 21–22).

Francisco always remained on warm emotional terms with his mother while growing increasingly distant and aloof toward an uncaring father: "With his father . . . he had as little as possible to do, and the father seems to have retained less affection for him than for his [two] brothers" (Trythall 1970, 25).

Francisco's early life was seriously jolted when the father abandoned the family in 1904, initially to serve in various naval posts and eventually to lead a philandering life:

[Francisco had] a rather unhappy childhood. His father . . . was a gay companion, a bibulous amorist with little taste for family life. He left home and Franco's mother, a "widow" during her husband's lifetime, was much given to hours of solitary weeping. . . . Since Franco's conscious childhood was spent under his mother's care and within earshot of whispered admonitions about his absent father, it is likely that these were the roots of his later puritanism. (Crozier 1967, 33–34; cf. Trythall 1970, 25)

Another biographer characterizes Nicólas Franco as "a rake," who left his wife "increasingly on her own in bringing up the children while her husband's behavior shamed her and scandalized the neighbors. . . . [Nicólas] never served in Ferrol again nor did he ever rejoin his wife" (Trythall 1970, 22, 25).

Later in the same year (1904) Francisco was sent to the Naval Preparatory Academy in anticipation of "the naval career which had always been planned for him" (Trythall 1970, 23). There, "Francisco grew up amidst the so-called 'Generation of 1898': a convenient historical tag for the reaction which affected Spain as a result of the catastrophic war with America" (Lloyd 1969, 21; cf. Crozier 1967, 13–14, 32). Defeat in the Spanish-American War had delivered a serious military and psychological blow to the Spanish navy, leading to serious recessions, contractions, and cutbacks. "Unable therefore to enter the Navy, Francisco decided [in 1907] to sit the examinations for the Infantry Academy at Toledo, for this was admitting as many cadets as ever" (Trythall 1970, 24–25; cf. Lloyd 1969, 23). Having successfully completed the Toledo Academy, Franco was commissioned in the Spanish army in August 1910.

The Toledo Academy also served to fan Franco's ultranationalism: "Toledo had anchored in him the uncomplicated patriotism of his childhood, and fostered hope that Spain might again enjoy her rightful prominence in the world.... The Galician is famous as a migrant, and Francisco Franco longed to escape from his family history and to prove himself in the service of Spain" (Trythall 1970, 26–27).

Franco's military career was meteoric. A series of rapid promotions followed, culminating in his appointment as brigadier general in 1926 at age thirty-two, the youngest of his rank in the Spanish army.

Military advance gave fresh urgency to Franco's ultranationalism, puritanism, and sense of duty. As early as 1928, he drafted Ten Commandments for the aspiring soldier—most likely patterned after Mussolini's Decalogue for the Italian soldier:

First Commandment: Love your country and be faithful to your King.

Second Commandment: Cultivate a great military spirit.

Third Commandment: Be chivalrous in spirit.

Fourth Commandment: Carry out your duties faithfully and precisely.

Fifth Commandment: Never grumble and do not tolerate the grumbles of others.

Sixth Commandment: See to it that you are loved by your inferiors and appreciated by your superiors.

Seventh Commandment: Be ready to volunteer for any sacrifice by asking—and wishing—to be used on occasions when the risks and the fatigue are greatest.

Eighth Commandment: Be a good comrade.

Ninth Commandment: Develop a love of responsibility and decision.

Tenth Commandment: Show courage and abnegation.

(quoted in Crozier 1967, 94)

MUSTAFA KEMAL ATATÜRK (NÉ MUSTAFA), 1881–1938

Mustafa Kemal Atatürk's military career was situational in many ways. His life spanned the chaos and turbulence that surrounded the decline and fall of the Ottoman Empire. Intensely nationalistic and acutely aware of his identity as a Turk, he carved out of the disintegrating empire the modern republic that is Turkey. Intensely anti-Islam (even though he was born a Muslim), he set out to Westernize and glorify his beloved country. This design has been attributed to the grandiose, narcissistic, and oedipal personality that he developed in childhood and beyond.

Mustafa Kemal Atatürk was born in Salonika, a provincial center of the Ottoman Empire, in 1881. He was named simply "Mustafa," one of the prophet Mohammad's two hundred names. (Not until the mid-1930s did it become legally necessary to have a surname in Turkey.) "Kemal," meaning perfection, was a name given him by a military superior while in the Military Cadet School (1892–1896), in recognition of his facility with mathematics and history. "Atatürk," meaning "Father Turk," was an accolade he preempted for himself in 1934, in the course of his founding and presidency (1923–1938) of the modern republic of Turkey. From this point on, he called himself Kemal Atatürk, dropping Mustafa altogether and reflecting his contempt for Islam as a primitive, ignorant, and backward religion.

Mustafa's father, Ali Reza, was a minor customs official who had difficulty looking after a family of five. Mustafa had a difficult childhood and a stormy home life. Rebellious from early on, he was expelled from grammar school for questioning authority and causing disturbance (Brock 1954, 7–8). Nonetheless, he was determined that "I shall *be* somebody!" (quoted in Brock 1954, 6; emphasis in original).

Mustafa's father and mother (Zübeyde) clashed over the child's schooling. The mother, a devout Muslim, wanted Mustafa to go to clerical school and follow a career as an Islamic teacher. The father, more liberal, wanted Mustafa to go to a secular, Western school. Ali Reza died in 1889, when Mustafa was seven years old, leaving the situation unresolved.

Meanwhile, Mustafa had become extremely impressed with a neighborhood boy, Ahmad, who attended the Military Cadet School in Salonika and "flaunted its uniform" (Kinross 1965, 13; cf. Volkan and Itzkowitz 1984, 34–35). With the help of Ahmad's father, an army major, Mustafa entered the military school in 1892, at age twelve. He consoled his mother by reminding her that "his father had presented him with a sword at birth and had hung it on the wall above his cradle. That could only mean that he had wished him to be a soldier. 'I was born a soldier,' he said, striking an attitude of heroism, 'I shall die as a soldier'" (Kinross 1965, 14).

In 1895 Mustafa entered the Senior Military School in Monastir in Turkish Macedonia. From 1899 to 1902 he attended the War College in Constantinople (now Istanbul). In 1902 Mustafa entered the General Staff School in Harbiyeh, graduating in 1905 at age twenty-four. By 1915 he had been promoted to the rank of general. In the course of his military career, Mustafa made a point of studying not only Napoleon, Clausewitz, and Moltke, but also Voltaire, Rousseau, J. S. Mill, and Hobbes—all as preparation for establishing a Westernized Turkish Republic.

The most exhaustive psychohistorical study of Atatürk has been undertaken by Volkan and Itzkowitz (1984), revolving around the themes of grandiosity, narcissism, and oedipal complex. Volkan and Itzkowitz summarize their findings in the following terms:

> We came early to the conclusion that Atatürk had an inflated and grandiose self-concept, basing this on the way others described him, but also—and more significantly—on his own delineation of his personality organization. He believed he was a unique man above all others and endowed with the right to assert his will. . . . Our clinical studies indicate that people with this kind of orientation . . . may have what is technically known as a narcissistic personality organization. This is a pathological response to psychological deprivation and trauma suffered when a child prematurely builds a sense of identity, and it involves a precocious and exaggerated sense of autonomy and omnipotence. . . .
>
> Clinical research . . . indicates that the adult with a grandiose self-concept had a special kind of early relationship with his mother. It often becomes clear that the mother was cold and nongiving toward her child and left him emotionally hungry and inclined to develop a self-concept that made him rise above hurts by being grandiose. At the same time she perceived him somehow as being "special," perhaps seeing him as an ornamental plaything or the savior of the family's name, and she regarded him as valuable to her in ways that she took pains to reinforce. . . .
>
> Those who in childhood develop an inflated self-concept (grandiose self) grow up to be excessively self-engrossed, with grandiose fantasies accompanied by an overdependence upon acclaim and an insatiable need to attain brilliance, power, and beauty. (Volkan and Itzkowitz 1984, xxiii–xxv)

Indeed, Atatürk sought to become "a mighty, godlike, and immortal leader, reflecting the defensive adaptation of his childhood in the world of public affairs" (Volkan and Itzkowitz 1984, xxv). In adulthood, he substituted the Turkish nation for his mother (Volkan and Itzkowitz 1984, 358 et passim).

SIMÓN BOLÍVAR, 1783–1830

For all of Simón Bolívar's life, Venezuela—indeed, all of Spanish America—was in a state of total turmoil: military, political, economic, social, and psychological. Spain was beginning to lose its grip on her colonies. The American and French Revolutions heralded a new age of liberty and equality for the peoples of Spanish America. The Napoleonic invasion of 1808–1812 further weakened Spain and fanned the fires of nationalism throughout the Spanish-American colonies.

Simón Bolívar was born in Caracas on July 24, 1783. Both Don Juan Vicente Bolívar and Doña María Palacios came from ancient aristocratic families. As fate would have it, however, the father died when Simón was three years old, and the mother followed him six years later. These deaths set in motion a serious line of love deprivation which had personal and political consequences for Simón Bolívar. On a personal level, Bolívar compensated by considering himself a Don Juan and by consorting with a galaxy of women in Europe and Spanish America (Masur 1948, 57 et passim; Ludwig 1942, passim). On a political level, he transferred his parental love onto nationalist sentiments and a life-long devotion to free the countries of Spanish America. He wrote: "Our life is nothing but the essence of our poor country" (quoted in Masur 1948, 32). Political and nationalist success, in turn, ignited in Bolívar vanity, egomania, and a quest for personal glory.

Upon his parents' deaths, young Simón was placed in the guardianship of his maternal uncle, Esteban Palacios, who in turn placed the boy with tutors and priests to be educated. During this period, Simón's only source of emotional support was his black nurse Hipolita. As an adult, he wrote to a sister: "I am sending you a letter from my mother, Hipolita, so that you will give her everything she wishes, and so that you will treat her as though she were your own mother. She nourished my life. I know no other parent than her" (quoted in Masur 1948, 31; cf. Worcester 1977, 9).

Between 1797 and 1799 (age fourteen to sixteen), at Uncle Palacios's urging, Bolívar joined the Corps of Cadets of Aragua Militia (which ironically his father had founded), attaining the rank of lieutenant. This is the extent of Bolívar's formal military training; he learned most everything else on the battlefield.

In 1799 Bolívar was sent to Spain in order to complete his general education, where he stayed for three years. In 1802 he married the daughter of a Spanish nobleman, María Teresa Rodríguez de Toro, with whom he returned to Caracas. Unhappily, María Teresa died of yellow fever less than a year later. Once again, at age nineteen, Bolívar found himself "bereft" (Masur 1948, 44). He began to console himself in the company of a large constellation of women. "To work he had to love, or rather make love, for Bolívar never really loved any woman" (Masur 1948, 249).

In 1804 Bolívar traveled to Paris where he personally witnessed the self-coronation of his long-time hero, Napoleon. The event temporarily disillusioned Bolívar, but in the long run he never wavered in his admiration for and emulation of Napoleon (in addition to Alexander the Great and Julius Caesar) (Worcester 1977, 205).

While in Paris, Bolívar met Alexander von Humboldt, the great Prussian scientist who had just completed a five-year expedition of Spanish-American countries. Humboldt told Bolívar he had found Spanish America ripe for independence, but that men of vision and determination were lacking. Humboldt challenged Bolívar to lead the liberation movement. Living a life of privilege and luxury, Bolívar felt "a sense of shame of his own useless existence" (Masur 1948, 55). His imagination inflamed, he accepted Humboldt's challenge.

In 1805 Bolívar met Humboldt again, this time in Rome, where he was reminded of the challenge. Going to Monte Sacro (Holy Mountain), accompanied by his long-time tutor, Simón Rodríquez, Bolívar knelt and vowed to liberate Venezuela from Spanish domination. (Rodríquez had systematically educated Bolívar in the ideas of the English, French, and American revolutions, as well as in the specific teachings of Rousseau, Voltaire, Montesquieu, Helvetius, Condillac, Locke, Hobbes, and others.)

Visiting the United States in 1806, Bolívar returned to Venezuela a year later. From that point on, "His one inspiration in word and deed was the freedom of [South] America. He pursued this ideal as Don Quixote pursued the ideal of knighthood.... The idea of freedom made him a prophet, as it made him an orator, an actor, a thinker" (Masur 1948, 260). Within two decades, Bolívar had achieved the independence not only of Venezuela but also of Bolivia, Colombia, Ecuador, and Peru. (Bolivia, needless to say, was named after Bolívar.)

As Bolívar's successes mounted, so did his vanity and egomania. He characterized life as "the desert of egoism" (quoted in Masur 1948, 44). His love for fame and glory became unquenchable. He claimed: "I am the son of war. War is my element—danger, my glory" (quoted in Worcester 1977, 63; cf. Masur 1948, 255). And again: "I want to fight for my glory at the cost even of the whole world" (quoted in Worcester 1977, 135).

Ambitious and self-centered, Bolívar came to believe that he had been chosen by God to liberate South America from Spain. Accordingly, he embraced the title of "Liberator" with unabashed enthusiasm: "The title of Liberator is beyond any reward ever offered to human pride" (quoted in Worcester 1977, 160). And again: "the position of Liberator is more sublime than a throne" (quoted in Worcester 1977, 170).

No longer did Bolívar want to be Alexander, Caesar, or Napoleon. Coming directly from God, he believed, the title of Liberator was higher than king or emperor. Bolívar loved to hear the people of Spanish America attending mass recite the following verse:

> O, Lord, all good things come from thee.
> Thou has given us Bolívar.
> Glory to Thee, great God!

What a man is he, O, heaven,
Who by thy hand with love and skill is crowned.
The future he knows as well,
As though time did his voice obey.
(quoted in Masur 1948, 565)

JOSIP BROZ TITO (NÉ JOSIP BROZ), 1892–1980

Josip Broz Tito was born Josip Broz. "Tito" is a nom de guerre he adopted in 1934 because of its literary affinity for his native district of Zagorje (Dediger 1953, 88; Auty 1970, 86).

Josip experienced relative deprivation in childhood, as a result of which he sought social advancement in various endeavors, including the military. His fierce nationalism was ignited by the harsh treatment Croats received under Austro-Hungarian rule.

Born to a poor peasant family, seventh of fifteen children, Josip's initial motivation for occupational, political, and military activity was to break through the cycle of poverty and hardship:

The impression that one gets of Tito during this period [1892–1920] is of a young boy desperate to get away from the hard life of his native village. He dreams of appearances: waiters and non-commissioned officers wear smart clothes, and those occupations consequently appealed to him. His aim in life is to learn a craft that will enable him to earn a better living in urban industries. The pattern is set. He works long enough to send money home at first, and to buy himself a suit, before he is off again in search of a better job and to see the world. Military service then opens the prospect of army life, at which he is successful in peace and war. (Pavlowitch 1992, 14)

The principal outline of Tito's career can be summarized readily enough. Between 1907 and 1913 he worked at a series of menial jobs. In 1913 he joined the Austro-Hungarian army, registering in a school for noncommissioned officers. He performed well, and in 1914 he became the youngest corporal and then the youngest sergeant in his regiment. During World War I he fought on a variety of fronts. In the interwar period he became a functionary—and then an official—of the Yugoslav Communist Party. During World War II he launched his legendary partisan warfare throughout Yugoslavia. In 1945 he assumed the rank of marshal. Given these circumstances, needless to say, Tito (like Bolívar) skipped the usual trajectory of a standard military career.

Tito said of the difficulties of his childhood:

A hard life awaited my parents. Fifteen acres of land, which dwindled as my father's debts came due, were insufficient to feed the family [of seventeen]. When the debts became intolerable, the soft and good-natured Franjo [Josip's father] gave it up and took to drinking, and the whole family burden fell upon my mother, an energetic woman, proud and religious. . . .

My childhood was difficult. There were many children in the family and it was no easy matter to look after them. Often there was not enough bread, and my mother was driven to lock the larder while we children received what she considered she could give us, and not what we could eat. In January my father had to buy cornmeal bread because we could not afford wheat. We children often took advantage of the visit of relatives to beg a slice of bread more than the ration we had eaten. My mother, a proud woman, would not refuse us before relatives. But after they went there was scolding and even occasional whipping. (quoted in Dediger 1953, 13–14)

In 1904, at age twelve, Josip came under the influence of a cousin, Jurica Broz, a staff-sergeant on leave from the Austro-Hungarian army. Jurica told Josip about the adventures of military life and about the possibilities of leaving his home village of Kumrovec.

In 1907 Josip headed for the garrison town of Sisak in search of fortune and adventure. Once in Sisak, Tito worked for a time as a waiter. He then apprenticed as—and became—a locksmith and a mechanic. He came under the influence of trade unionism, socialism, and Marxism. In 1910 he joined the Metal Workers' Union, which automatically made him a member of the Social Democratic Party of Austria. Having been disillusioned with Sisak, Tito traveled to Slovenia, Zagreb, Vienna, and Munich, working at menial jobs. In 1913 he joined the Austro-Hungarian Army, which turned out to be a radicalizing agent.

Service with the Austro-Hungarian army was to bring experiences that changed Josip Broz from a conformist into a potential rebel and social outcast. . . . Broz saw a social order overthrown during the Russian revolution, and people at the bottom of the ladder, like himself, suddenly elevated to power. It was a heady experience. His own change into a political rebel was not a sudden conversion, but the change when complete was a permanent one, deeply rooted in his personality and linked with his childhood experiences. It had complex causes in which ambition, personal disappointment, idealism, and human commitment all played their part. (Auty 1970, 29–30)

Tito formally joined the Communist Party in 1923, five years after Yugoslavia became independent. "His explanation for why he became a communist lay in the injustices of society, the poverty and oppression he had seen as a child, which were being continued in the bourgeois-ruled society of the new Kingdom of Yugoslavia" (Auty 1970, 3).

Tito's experiences helped fan the nationalist spirit that had been simmering since childhood. Making a searing impression on Tito from a very early age were the oppression and exploitation that Croats suffered under Hungarian rule, the deliberate Hungarian policy of keeping Croatia poor and undeveloped, the pervasive poverty, and the heavy burden of taxation. "A vivid memory of Tito's childhood was of a revolt in the Zagorje villages in 1903 when he was eleven years old, against increased taxes. Peasants . . . showed their feeling by pulling down the Hungarian flag from a local railway station" (Auty 1970, 7).

A biographer writes:

> Stories of folk history told of the terrible experiences of Zagorje peasants under centuries of feudal oppression when the majority of Croats had been serfs without legal rights under Hungarian law. . . . The dramatic stories of their own past were told and retold. The heroes of these stories—in which good was identified with their own people, the poor and the oppressed, and evil with the ruling class—were those who showed independence, defied authority, and usually died horribly. The drama was heightened by the fact that the stories were based on historical events and local associations. (Auty 1970, 7)

Tito's nationalism reached a peak of intensity during World War II. In 1941, following the Axis invasion of Yugoslavia, Tito issued a statement:

> It proclaimed that the main purpose of the Partisan detachments was the liberation of the peoples of Yugoslavia from the occupation forces, and the struggle against those who were assisting them in oppressing and terrorizing our people. Tito stressed that the Partisan detachments were called National Liberation Detachments because they were fighting formations not of any political party and group—not even the Communist Party, although the Communists were in the forefront of the struggle—but were the fighting forces of the people of Yugoslavia and should therefore include all patriots, whatever their views, who were capable of waging an armed struggle against the invaders. (Dediger 1953, 156–57)

Following World War II, Tito's nationalism proved sufficiently strong even to defy Stalin, leading to the "expulsion" of Yugoslavia from the world communist movement in 1948.

ed
Synthesis

Chapter 9

Military Leaders in Theoretical Perspective

In this book we have developed and applied an interactional theory to forty-five major military figures across time and space. It remains for us to synthesize our findings, offer some probabilistic statements concerning the chief variables that may propel persons toward military careers, and present some closing remarks. We should reiterate that since we do not have a representative sample of military leaders, the generalizations that follow apply only to the forty-five individuals studied in this work.

SOCIODEMOGRAPHIC FINDINGS

The great majority of the hypotheses and propositions presented in Chapter 1 relative to the sociodemographic, experiential, ideological, and attitudinal attributes of military leaders are supported by the data.

Military leaders are mature men (over age forty-five) upon reaching the highest rank. Pre-twentieth-century leaders tend to be somewhat younger, perhaps due to the less complex nature of warfare and command at earlier times. Military leaders are exposed to military ideology—and they participate in combat—early in life, sometimes in their teens.

Although a majority of military leaders are rural born, *all* military elites develop early and sustained exposure to urban cultures. As centers of values and skills, urban cultures are the seedbeds of all types of leadership.

Military leaders come from the main ethnic groupings of their societies. Their religious backgrounds and orientations are also of the mainstream variety.

Military leaders are well educated, with an overwhelming majority graduating from military academies.

Military leaders are overwhelmingly committed to the military as their sole occupation. The primary occupations of the fathers of military leaders are primarily the military and the professions, with a smattering of other pursuits.

With rare exceptions, military leaders participate in legal organizations and activities. The few who participate in radical or revolutionary action compile records of arrest, imprisonment, or exile.

Military leaders adopt conservative and indigenous political ideologies. Radical and foreign ideologies are extremely rare.

Military leaders have a dualistic view of human nature, distinguishing between good (their own people) and evil (other peoples). Military leaders have a uniformly positive view of their own countries, while their attitude toward the international community is dualistic, distinguishing friends and enemies.

As anticipated in Chapter 1, some of our findings are "situational" in nature—that is, attributes over which the leaders have little or no control. These include: birthplace, ethnicity and religion, social class, father's occupation, early exposure to military ideology and military action. These attributes lend additional credence to our interactional theory of military leaders.

In any event, given their education, occupation, and status, military leaders are strategically situated in their societies, commanding ready access to positions of power and authority.

Only three of the hypotheses advanced in Chapter 1 are not supported by the data.

First, the socioeconomic status of military leaders has undergone significant changes in recent times. More precisely, in the twentieth century, upper class representation has significantly dropped while middle and lower class representation has sharply risen. Significant numbers of fathers of military leaders have lower class occupations. In other words, the military has increasingly become an avenue of upward social mobility. While this finding has long held for the military rank and file, its direct replication for high-level leaders is rather surprising.

Second, military leaders have many siblings, but contrary to expectations, middle children are overrepresented among military elites, while youngest, oldest, and only children are underrepresented. Our earlier findings concerning the relative inactivism of middle children do not hold for military leaders. Alternatively, the secure life of the military—particularly in peacetime—may be attractive to certain types of personalities.

Third and finally, military leaders are not as cosmopolitan as we anticipated. Few speak foreign languages; few travel abroad before reaching the highest rank; few cultivate foreign contacts.

These exceptions notwithstanding, military leaders constitute a remarkably homogeneous group of individuals.

PSYCHOLOGICAL FINDINGS

The six motivational and psychological dynamics put forward in Chapter 1 and analyzed in Part III are supported, in varying degrees, by the evidence available to us. It is now time to summarize and synthesize our findings. We list the six dynamics below, indicate the individuals associated with each dynamic, and (to avoid redundancy) offer sample commentaries. The entire picture is summarized in Table 9.1, wherein "X" signifies the presence of an attribute.

Nationalism

To some degree or another, all of the forty-five military leaders are, almost by definition, nationalists: they are intensely committed to their nations, and they seek to enhance the prestige, status, and power of their countries.

Imperialism

Sixteen of the forty-five leaders (35.5 percent) exhibit some form of imperialist design: Bonaparte, Burgoyne, Chennault, Dayan, de Gaulle, Ludendorff, MacArthur, Mahan, Moltke, Nelson, Pétain, Potemkin, Rommel, Tojo, Yamamoto, and Zhukov.

Mahan was a racist, a Social Darwinist, a believer in the white man's burden, and a champion of the Monroe Doctrine; he felt called upon to spread Anglo-Saxonism throughout the world. Napoleon's imperialism ignited the European wars of the nineteenth century. Yamamoto played a key role in planning the attack on Pearl Harbor. Tojo was a believer in Japanese racial supremacy; stressing the divine origin of his country, he believed that Japan was called upon to conquer the known world.

Relative Deprivation

Twenty military leaders (44 percent) experienced relative deprivation of various kinds: Atatürk, Bonaparte, Churchill, Clausewitz, Denikin, Eisenhower, Franco, Jackson, Jones, Lee, Ludendorff, Marshall, Moltke, Nelson, Pershing, Potemkin, Sherman, Tito, Yamamoto, and Zhukov.

For example, in the case of Pershing, during the panic of 1873 his father's mercantile business and landholdings were wiped out, making life very harsh for a family of eight. Eisenhower, the third of seven sons born to a desperately poor family, experienced scarcity and want. Chur-

Table 9.1
Psychological Attributes of Military Leaders

Leader	Attribute					
	Nationalism	Imperialism	Relative deprivation	Love deprivation	Marginality	Vanity
Atatürk	X		X	X		X
Bolívar	X			X		X
Bonaparte	X	X	X		X	X
Burgoyne	X	X			X	X
Chennault	X	X		X		X
Churchill	X		X			
Clausewitz	X		X		X	
Dayan	X	X		X		X
De Gaulle	X	X			X	X
Denikin	X		X			
Eisenhower	X		X			
Fairfax	X			X	X	
Foch	X					X
Franco	X		X			
Garibaldi	X					
Gordon	X					

126

Halsey	X				
Hindenburg	X				
Jackson	X	X	X		
Jones	X	X		X	
Kitchener	X		X	X	X
Lafayette	X		X	X	X
Lee	X	X	X	X	
Ludendorff	X	X			X
MacArthur	X				X
Mahan	X	X			
Marshall	X	X	X	X	
Moltke	X	X	X	X	
Montgomery	X	X	X		X
Mountbatten	X				X
Nelson	X		X		
Nimitz	X			X	X
Patton	X				X
Pershing	X				X

Table 9.1 (Continued)

	Attribute					
Leader	Nationalism	Imperialism	Relative deprivation	Love deprivation	Marginality	Vanity
Pétain	X	X		X		X
Potemkin	X	X	X	X		X
Rommel	X	X				
Santa Anna	X					X
Sherman	X		X			X
Tojo	X	X				
Tito	X		X			
Washington	X			X	X	
Wellesley	X				X	
Yamamoto	X	X	X			
Zhukov	X	X	X			X
Total	45	16	20	17	16	22

chill knew a similar situation, except that he had twelve siblings of whom seven survived to adulthood. Moltke's father was a member of the impecunious aristocracy; in addition, being in poor health, unstable, and temperamental, he had great difficulty looking after a family of ten. Ludendorff and Potemkin had similar experiences. Tito, born to a poor peasant family of fifteen children, lived a life of poverty and hardship.

Persons who experience relative deprivation typically perceive the military as a means of escaping the miseries of home and as an avenue of upward social mobility.

Love Deprivation

Seventeen military leaders (38 percent) experienced varieties of love deprivation: Atatürk, Bolívar, Chennault, Dayan, Fairfax, Jackson, Kitchener, Lafayette, Lee, Marshall, Moltke, Montgomery, Nelson, Pétain, Potemkin, Washington, and Wellesley.

In the case of Washington, his father died when the boy was eleven years old, and his relations with his mother always remained tense, strained, and distanced. Jackson's father died before the child was born, and his mother died when Jackson was fourteen. Montgomery had a passive and distant father and a tyrannical mother; as the fourth child and third son in a family of nine children, he felt alone, neglected, and cast aside. Lafayette's father was killed in battle when the boy was two years old; abandoning him until age eleven, his mother died two years later. Lafayette set out in search of a surrogate father, whom he eventually found in George Washington. Dayan had a rather stormy family life, and he became increasingly distant from—and disillusioned with—his absent father. Atatürk's father died when the boy was seven years old, and he had a cold and distant mother. Bolívar's father died when the boy was three; his mother died six years later.

Persons experiencing love deprivation substitute the military for the lost parent(s) and for the security and affection they missed at home.

Marginality

Sixteen military leaders (35.5 percent) are in one way or another marginal: Bonaparte, Burgoyne, Clausewitz, de Gaulle, Fairfax, Hindenburg, Jackson, Jones, Lafayette, Mahan, Marshall, Moltke, Nelson, Pershing, Washington, and Wellesley. Marginality may take a variety of forms: social, cultural, physical, and medical/psychological.

Social marginality. Napoleon never forgot his Corsican background; he was routinely ridiculed for speaking accented and ungrammatical French; even at the height of his power he was never fully accepted by French aristocracy, who called him "the Corsican upstart" and "the Corsican ogre." For Jackson, given his social background, he had difficulty finding social acceptability. Hindenburg was ashamed of his mother's middle class background, preferring instead to trace his noble paternal lineage to 1289.

Cultural marginality. Washington and Jackson never properly learned the English language. Clausewitz was shy, nervous, withdrawn, and awkward, and so was scorned and ridiculed by his fellow officers and treated as an outsider. Lafayette had an inferiority complex; felt out of place in the salons of Paris; and was routinely ridiculed by his father-in-law for lack of grace, polish, and wit. Moltke's experiences in a Danish military academy produced culture shock and an inferiority complex, leaving a lasting impression on the young boy. Pershing was variously called "Nigger," "Nigger Jack," and "Black Jack."

Physical marginality. Napoleon was pitifully short and thin; he had an awkward manner; and he was derisively called "the Little Corporal." Nelson was a sick and puny boy; he had a small stature; having lost an eye and an arm, he looked like a cripple. Wellesley had great difficulty competing physically and intellectually with his two elder brothers; his mother called him "an ugly boy" and "an awkward son." Fairfax was lanky, dark skinned, and dark haired, and so was nicknamed "Black Tom." Marshall was physically awkward as a child and a youngster. De Gaulle was ridiculed by schoolmates for his gangling frame and his prominent nose.

Medical/psychological marginality. Nelson and Clausewitz regularly experienced periods of depression and melancholy. Mahan developed mental depression from age sixteen onward, and a drinking problem from age twenty onward. Burgoyne was an illegitimate child.

Leaders who experience various kinds of marginality typically compensate by developing ambition, vanity, egotism, and a thirst for glory and heroism.

Vanity, Egotism, Narcissism

We have learned in our previous studies that a degree of vanity is a defining characteristic of almost any leader in any field. Of the forty-five military leaders, 22 (nearly half) are particularly noted for their vanity: Atatürk, Bolívar, Bonaparte, Burgoyne, Chennault, Dayan, de Gaulle, Foch, Jones, Kitchener, Lafayette, MacArthur, Mahan, Montgomery, Nelson, Patton, Pershing, Pétain, Potemkin, Santa Anna, Sherman, and Zhukov.

To avoid redundancy, we simply note that these leaders are variously vain, egotistical, narcissistic, megalomaniacal, conceited, arrogant, haughty, sarcastic, contemptuous, temperamental, mercurial, and seekers of glory and heroism. Accordingly, Napoleon crowned himself and claimed power as his mistress. Admiring Napoleon, Bolívar came to believe that he had been sent by God to liberate Spanish America. Santa Anna and Patton considered themselves to be Napoleons incarnate. MacArthur saw himself as a sovereign. Mustafa Kemal preempted for himself the title Atatürk. Pétain and de Gaulle viewed themselves as France incarnate.

Strikingly, we came across a military leader who was totally devoid of vanity, desire for personal recognition, or search for glory: Giuseppe Garibaldi. In 1860, he declined King Victor Emmanuel's offer of the rank of major general, a princely title, a large pension, a castle, and the nomination of his son Menotti as the king's aide-de-camp.

Addendum: Unanticipated Findings

In exploring the lives of the forty-five military leaders, we came across two sets of unanticipated findings: (1) religiosity, mysticism, superstition; and (2) facility with mathematics.

Religiosity, mysticism, superstition. These attributes are found in 10 of the forty-five leaders (22 percent): Bonaparte, Foch, Gordon, Halsey, MacArthur, Mahan, Montgomery, Nelson, Patton, and Wellesley.

Napoleon routinely made the sign of the cross as a means of protecting himself. Just before the Battle of Waterloo, he insisted that fortune had abandoned him.

Nelson had a death wish: he was convinced that some day a bullet would strike him. As it turned out, his mysticism came to pass: he was killed by a French sharpshooter at the Battle of Trafalgar.

Wellesley was fatalistic and a strong believer in the intervention of the Providence. He considered the British army to be God's Almighty Army.

Gordon inherited from his mother a strong evangelical streak. While at military academy, he came under the influence of a superior who was also deeply religious. The military and the religious merged. Gordon was a mystic, a fatalist, and a fanatic. He also developed a death wish.

Montgomery had an element of mysticism about him. In his 1960 book on military leadership, he included an epilogue titled "In My Garden," where, falling asleep, he carried on a conversation with the ghost of his beloved father.

Foch was profoundly religious: he thought of Catholicism as the mainspring of all behavior. He believed spiritual values to be supreme, taking precedence over all else.

MacArthur, Mahan, and Patton—all Episcopalians—saw themselves as the defenders of Christendom and as agents of Christianity. Patton believed he had been a soldier in previous lives, and that he would be born again—as a soldier.

Halsey was superstitious in general and about the number thirteen in particular.

Mathematical facility. Ten of the 45 leaders (22 percent) are mathematically inclined: Atatürk, Bonaparte, Clausewitz, Denikin, Foch, Lee, Ludendorff, Pétain, Rommel, and Washington. Napoleon, who is most explicit on this point, wrote in his diary for January 29, 1818:

> To be a good general a man must know mathematics; it is of daily help in straightening one's ideas. Perhaps I owe my success to my mathematical conceptions; a general must never imagine things, that is the most fatal of all. My great talent, the thing that marks me most, is that I see things clearly; it is the same with my eloquence, for I can distinguish what is essential in a question from every angle. (Bonaparte 1910, 499)

Napoleon makes mathematical ability almost an inherent psychological trait or phenomenon.

Summary

With four exceptions, the forty-five military leaders possess the six psychological attributes in various combinations. Nelson is the only individual to meet all six variables. Bonaparte, Moltke, and Potemkin each possess five attributes. Most leaders meet two, three, or four variables. The four exceptions—Garibaldi, Gordon, Halsey, and Nimitz—meet only the variable of nationalism.

The most pervasive psychological dynamic is nationalism, followed by vanity, relative deprivation, love deprivation, imperialism, and marginality.

Two unanticipated findings characterize some of the forty-five military leaders.

SITUATIONAL FINDINGS

We have maintained throughout this work that, given the appropriate social backgrounds and psychological attributes, men are propelled into military careers as a result of "situational" variables. Astonishingly, this turns out to be the case for all the forty-five leaders. Four types of situations—at times overlapping—emerge: (1) birthplace, (2) family influences, (3) national crises, (4) luck or chance. Let us consider the evidence.

Birthplace

Nine military leaders (20 percent) are born in places that are conducive to military careers: Franco, Gordon, Hindenburg, Jones, Lee, MacArthur, Mahan, Pétain, and Yamamoto.

Franco, born in El Ferrol, a navy town, became immersed in naval matters from an early age. Gordon was born in the garrison town of Woolwich, where his father eventually rose to become a lieutenant general. Hindenburg was born at Posen, a Prussian military fortress; his childhood was spent going from one military post to another, as his father's career necessitated.

Jones, born in coastal Scotland, was deeply influenced by the sights and sounds of the sea and by the stories of seamen. Lee was born in Alexandria, near Mount Vernon; from childhood he was totally immersed in the glorification of the Revolutionary War, George Washington, and military service. MacArthur, born on an army post near Little Rock, Arkansas, spent much of his childhood and youth in various military posts, as his father's career required; he developed an early fascination with the military.

Mahan was born and raised on the grounds of the U.S. Military Academy at West Point, where his father was a senior professor. Pétain, born in a garrison town, became greatly impressed with the ways of the military and resolved to become a soldier in his teens. Yamamoto was born to a family of samurai in a city of seafaring men, making a naval career a natural course of action.

Family Influences

Fully twenty-nine individuals (64 percent) were either born directly to military families or chose a military career under the influence of a family member or a close family friend. Born into military families—and carrying on the family tradition—are twenty-two leaders:[1] Burgoyne, Churchill, Clausewitz, de Gaulle, Denikin, Fairfax, Foch, Franco, Gordon, Halsey, Hindenburg, Kitchener, Lafayette, MacArthur, Mahan, Marshall, Moltke, Mountbatten, Patton, Potemkin, Tojo, and Yamamoto.

Seven of the twenty-nine leaders chose a military career under the influence of a family member or a close family friend: Lee, Napoleon, Nelson, Pétain, Tito, Washington, and Wellesley.

Lee came under the guardianship of Senator Thomas Ewing, a close family friend, following his father's death when the boy was nine years old. Ewing arranged for Lee to enter West Point at age eighteen.

Napoleon showed a precocious interest in military matters, an interest that was thoroughly cultivated by his father, Carlo. At age ten, Carlo entered Napoleon in the military academy at Brienne.

Nelson came under the direct influence of his maternal uncle, Captain Maurice Suckling, following his mother's death when the boy was nine years old. When Nelson was twelve, Suckling registered the boy as a midshipman.

Pétain was inspired to join the French Army by the stories of a granduncle who had served under Napoleon. Tito, persuaded by a cousin, joined the Austro-Hungarian army. Washington emulated his idol and older half-brother Lawrence in every respect, including his military career.

Wellesley's father died when the boy was twelve years old, leaving him at the mercy of a cold and austere mother. Considering her son useless for any other pursuit, she dispatched him to a military academy.

National Crises

It is likely that many individuals steer toward the military as a response to national crises or emergencies. We have direct evidence for ten individuals (22 percent): Atatürk, Bolívar, Dayan, Jackson, Jones, Lee, Ludendorff, Rommel, Santa Anna, and Tito.

Atatürk's life spanned the chaos and the turbulence that surrounded the decline and fall of the Ottoman Empire; he lived under conditions of continuing crises.

For all of Bolívar's life, Spanish America was in a state of total turmoil: military, political, economic, social, and psychological. The American and the French revolutions heralded a new age of liberty and equality for all peoples. The Napoleonic invasion further weakened Spain and fanned the fires of nationalism throughout Spanish America.

Given Dayan's life span, given the aftermath of the two world wars, given the turbulence of the times in Palestine, and given the Jewish quest for a homeland, every Jewish person was intensely involved in nationalism (or Zionism), and every Jewish life had a military component.

Jackson joined the Continental Army in 1780, in the midst of the American Revolution, at age thirteen. Jones's life coincided with Britain's imperial expansion, when many young men left home in search of glory, riches, and adventure.

Lee, as we have seen, lived in the afterglow of the Revolutionary War, Mount Vernon, George Washington, and military service. In addition, in 1814 (when he was seven years old), the British captured his hometown of Alexandria, leaving a lasting impression on the young boy.

Ludendorff's military career began in the era of Bismarck and ended in the age of Hitler. He was a child of perpetual crises.

Rommel's dream to become a soldier (even in his teens) is not surprising, given the aftermath of the Franco-Prussian war, given Germany's quest for global supremacy, and given the constant glorification of the Fatherland.

Santa Anna was a child of his chaotic environment. His life and career reflect Mexico's turbulent history in the nineteenth century: chaos, revolution, continual warfare.

Tito's life and career coincided with the turmoil and chaos that surrounded the collapse of the Austro-Hungarian Empire. He, too, was a child of crises.

Luck or Chance

The military careers of three American leaders—Eisenhower, Nimitz, and Pershing—were due to luck or chance.

Eisenhower, having graduated from high school at age nineteen, had no sense of direction or purpose. A friend told him about the service academies. With the intercession of a senator, Eisenhower entered West Point two years later.

For Nimitz, a chance encounter with some soldiers created instant fascination with the military. A congressman offered him an opening at Annapolis, which Nimitz had not even heard of at the time.

Pershing, bored and unemployed, accidentally came across a newspaper announcement for a competitive examination for West Point, which he entered at age twenty-two.

Summary

The most compelling situational dynamics are family influences, followed distantly by national crises and birthplace, and very remotely by luck or chance.

BECOMING MILITARY LEADERS

Throughout this study we have identified a series of sociodemographic, experiential, attitudinal, motivational, psychological, and situational findings about military leaders. Is it possible to employ our key findings to construct a probabilistic, composite profile of individuals in a given society who are likely to become military leaders?

Before proceeding, we should note a major caveat. Since we have relied on a nonrandom, purposive sample of military leaders, we can develop—consistent with canons of contemporary social science—only descriptive-historical (or "modal") models of leader traits and attributes. Accordingly, we use the word "probabilistic" advisedly and with due caution.

Whether or not a person becomes a military leader depends on the possession or acquisition of certain key attributes and experiences. Distilled from the findings presented in this work, these variables are as follows.

1. The person is a native-born male.
2. The person is born to a military family.

3. The person is born in a military town or garrison.
4. The person experiences relative deprivation or love deprivation.
5. The person is vain and egotistical.
6. The person is a nationalist or an imperialist.

Needless to say, one's chances of becoming a military leader are maximized if one fits all the six variables; the chances are minimized to the extent that one does not fit the variables. What may not be immediately apparent is that the foregoing variables echo our interactional theory of military leaders. The first three are strictly situational in that one has no control over them; the last three refer to personal attributes and experiences that propel one toward the military.

CLOSING REMARKS

We have developed and applied an interactional theory to forty-five major military leaders across time and space, stressing the interplay of sociodemographic variables, psychological dynamics, and situational factors. With minor exceptions, the theory has held up quite well against both quantitative and qualitative data. We hope to have shed light on the social backgrounds of military leaders and on the motivational, psychological, and situational dynamics that propel them toward military careers.

The integration of the interactional theory with our previous work on loyalist and revolutionary elites—and the development of a unified theory of leadership—remain tasks to be accomplished in future works.

NOTE

1. In Chapter 2—Table 2.4(C)—we reported only sixteen fathers who had military occupations; that figure excludes six fathers who had a *combination* of occupations, including the military.

Appendix: General Code Sheet

Column(s)	Code	Variables and Response Values
1-2	____	Country (Name) _____
3-4	____	Leader (Name) _____
5	____	Type of Country _____

        0   Developed   ⎰ Give % of labor
        1   Semideveloped   ⎱ force in
        2   Undeveloped   agriculture
        9   Unknown

6-7	____	Date of Birth
		_____(exact)
8-9	____	Leader's Age on Reaching Highest Rank ____(exact)

        93   0-15
        94   16-19
        95   20-24
        96   25-34
        97   35-44
        98   45-64
        99   65 and over

10-11	____	Approximate Age When First Exposed to Military Ideology _____(exact)

        93   0-15
        94   16-19
        95   20-24
        96   25-34
        97   35-44
        98   45-64
        99   65 and over
        09   Unknown

Column(s)	Code	Variables and Response Values

Column(s): 12-13   **Code:** ____

Approximate Age When First Took Part in Actual Combat _____(exact)

    93 0-15
    94 16-19
    95 20-24
    96 25-34
    97 35-44
    98 45-64
    99 65 and over
    09 Unknown

14   ____

Sex _____

    0 Male
    1 Female
    9 Unknown

15   ____

Birthplace _____

    0 Urban
    1 Rural
    9 Unknown

16   ____

Exposure to Urban Life (if Born and Raised in Nonurban Areas) _____

    0 None
    1 little (less than one year)
    2 moderate (one to three years)
    3 extensive (four or more years)
    8 other (specify)
    9 unknown/not applicable

17   ____

Age at Which Exposed to Urban Life (if Born and Raised in Nonurban Areas) _____

    0 under 15
    1 15-19
    2 20-24
    3 25-29
    4 30-34
    5 35-39
    6 40-44
    7 45-49
    8 over 50
    9 unknown/not applicable

Column(s)	Code	Variables and Response Values
18	___	Father's Birthplace _____

        0   same as son
        1   urban, different than son
        2   rural, different than son
        7   foreign
        8   other (specify)
        9   unknown

19   ___   Religious Background _____

        0   Atheist
        1   Christian: Protestant
        2   Christian: Catholic
        3   Christian: Other
        4   Jewish
        5   Muslim
        6   Hindu
        7   Buddhist
        8   Other (specify)
        9   Unknown

20   ___   Religious Orientation _____

        0   Atheist
        1   Christian: Protestant
        2   Christian: Catholic
        3   Christian: Other
        4   Jewish
        5   Muslim
        6   Hindu
        7   Buddhist
        8   Other (specify)
        9   Unknown

21   ___   Parent's Religion _____

        0   Same
        3   Different (specify)
        8   Other (specify)
        9   Unknown

Column(s)	Code	Variables and Response Values
22	___	Family Life: Number of Siblings _____

      0 none
      1 one
      2 two
      3 three
      4 four
      5 five
      6 six
      7 seven or more
      8 other (specify)
      9 unknown

| 23 | ___ | Family Life: Age Ranking Among Siblings _____ |

      0 only child
      1 youngest child
      2 middle child (specify)
      3 oldest child
      4 oldest son
      5 youngest son
      6 only son
      8 other (specify)
      9 unknown

| 24 | ___ | Family Life: Status of Leader _____ |

      0 legitimate
      1 illegitimate
      2 illegitimate, parents subsequently married
      8 other (specify)
      9 unknown

| 25 | ___ | Family Life: Status of Parents _____ |

      0 parents married throughout leader's childhood and beyond
      1 parents separated before leader left home
      2 parents separated after leader left home
      3 parents divorced before leader left home
      4 parents divorced after leader left home
      5 parents separated due to a death: before leader left home
      6 parents separated due to a death: after leader left home
      7 parents unmarried at time of leader's birth
      8 other (specify)
      9 unknown

Column(s)	Code	Variables and Response Values
26	___	Family Life: Character _____

        0 broken home
        1 relatively tranquil
        2 relatively stormy
        8 other (specify)
        9 unknown

| 27 | ___ | Ethnic Background _____ |

        0 main ethnic group
        1 ethnic minority: large (specify)
        2 ethnic minority: small (specify)
        8 other (specify)
        9 unknown

| 28 | ___ | Socioeconomic Status _____ |

        0 upper class
        1 middle class
        2 lower or working class
        8 other (specify)
        9 unknown

| 29 | ___ | (Blank) |
| 30 | ___ | Education: Type _____ |

        0 private: secular
        1 private: church related
        2 state: secular
        3 state: church related
        4 private and state: secular
        5 private and state: church related
        6 private and state: secular and church related
        8 other (specify)
        9 unknown

Column(s)	Code	Variables and Response Values
31	___	Education: Place _____

        0 domestic
        1 domestic, foreign sponsored or supported institution
        2 U.S.
        3 the former U.S.S.R.
        4 Europe other than the former U.S.S.R. (specify)
        5 Asia (specify)
        6 Africa (specify)
        7 Latin America (specify)
        8 other including combinations (specify)
        9 unknown

| 32-33 | _ | Education: Highest Level Attained _____ |

        15 none, through high school
        16 some college, to B.A.
        17 graduate or professional
        18 military
        19 no formal education
        09 unknown

| 34 | ___ | Education: Highest Level Attained, Characteristics _____ |

        0 private and domestic
        1 private and foreign (specify)
        2 state and domestic
        3 state and foreign (specify)
        8 other (specify)
        9 unknown/not applicable

| 35 | ___ | Education: Field _____ |

        0 professional (specify)
        4 social sciences, humanities, the arts
        7 military
        8 other (specify)
        9 unknown/not applicable

| 36-37 | ___ | Father's Education: Highest Level Attained _____ |

        15 none, through high school
        16 some college, to B.A.
        17 graduate or professional
        18 military
        19 no formal education
        09 unknown, not applicable

Column(s)	Code	Variables and Response Values
38	____	Father's Education: Highest Level Attained: Characteristics _____

        0   private and domestic
        1   private and foreign (specify)
        2   state and domestic
        3   state and foreign (specify)
        8   other (specify)
        9   unknown/not applicable

| 39 | ____ | Father's Education: Field _____ |

        0   professional
        4   social sciences, humanities, the arts
        7   military
        8   other (specify)
        9   unknown/not applicable

| 40 | ____ | Foreign Languages _____ |

        0   none
        1   one
        2   two
        3   three
        4   four
        5   five
        6   six or more
        8   other, including bilingual (specify)
        9   unknown

| 41 | ____ | Foreign Travel Before Attaining Highest Rank: Extent _____ |

        0   none
        1   little (one country)
        2   moderate (two-three countries)
        3   extensive (four or more countries)
        8   other (specify)
        9   unknown

| 42 | ____ | Foreign Travel Before Attaining Highest Rank: Duration _____ |

        0   none
        1   short (less than one year)
        2   moderate (one-three years)
        3   long (four or more years)
        8   other (specify)
        9   unknown

Column(s)	Code	Variables and Response Values
43	____	Foreign Travel Before Attaining Highest Rank: Place _____

        0   none
        1   Europe only
        2   U.S.A. only
        4   Asia only
        5   Africa only
        6   Latin America only
        7   combination (specify)
        8   other (specify)
        9   unknown

| 44 | ____ | Continuing Foreign Contacts _____ |

        0   none
        1   few (one-three)
        2   some (four-six)
        3   many (seven or more)
        8   other (specify)
        9   unknown

| 45-46 | ____ | Primary Occupation _____ |

        22   military
        23   professions
        24   professional revolutionary
        26   working class
        27   combination
        28   government official
        29   politician
        30   businessman
        31   landlord
        19   other (specify)
        09   unknown

| 47-48 | ____ | Father's Primary Occupation _____ |

        22   military
        23   professions
        24   professional revolutionary
        26   working class
        27   combination
        28   government official
        29   politician
        30   businessman
        31   landlord
        19   other (specify)
        09   unknown

Column(s)	Code	Variables and Response Values

Column(s): 49   Code: ____   Membership in Any Organizations: Extent _____

    0 none
    1 few (one-three)
    2 some (four-six)
    3 many (seven or more)
    9 unknown

50   ____   Membership in Political Organizations: Traditional or Legal _____

    0 no
    1 yes
    8 other (specify)
    9 unknown

51   ____   Membership in Political Organizations: Radical or Revolutionary _____

    0 no
    1 yes
    8 other (specify)
    9 unknown

52   ____   Membership in Trade Union Organizations _____

    0 no
    1 yes
    8 other (specify)
    9 unknown

53   ____   Membership in Professional Organizations _____

    0 no
    1 yes
    8 other (specify)
    9 unknown

54   ____   Membership in Other Organizations (specify) _____

    0 no
    1 yes
    8 other (specify)
    9 unknown

Column(s)	Code	Variables and Response Values
55-56	____	Record of Arrests, Imprisonment, Exile, etc. ____

        11 none
        12 some (one-three times)
        13 moderate (four-six times)
        14 frequent (seven or more)
        09 unknown

| 57-58 | ____ | Duration of Imprisonment or Exile (specify) ____ |

        22 none
        23 less than one year
        24 one to nine years
        25 ten or more years
        09 unknown

| 59 | ____ | Publications ____ |

        0 none
        1 few (one-three)
        2 some (four-six)
        3 many (seven or more)
        9 unknown

| 60-61 | ____ | Type of Ideology to Which Leader Subscribes ____ |

        24 none
        30 democratic
        31 Marxist-Leninist
        32 Nationalist/Marxist-Leninist (Marxist-Leninist-Castroite, Nationalist/Marxist-Leninist-Maoist)
        34 Nationalist/other (Nationalist, Nationalist-Socialist, democratic-nationalist)
        35 Leftist/other (utopian socialist, anarchist-socialist, Marxist-socialist, Marxist-anarchist, Trotskyite, Jacobin, Puritan-radicalist, democratic-socialist)
        36 Rightist/other (Fascist, Nazi, conservative monarchist, constitutional monarchist)
        37 other (republican) (specify)
        38 vacillating
        39 Conservative Nationalist
        09 unknown

| 62 | ____ | Origin of Ideology ____ |

        0 primarily indigenous
        1 primarily foreign
        2 primarily foreign but adapted to indigenous conditions
        8 other (specify)
        9 unknown

Column(s)	Code	Variables and Response Values
63	____	Personalities the Leader Admires or Identifies With _____

    0 none
    1 military (specify)
    2 scholar/writer/intellectual (specify)
    3 revolutionary (specify)
    4 philosopher (specify)
    5 national hero: own country (specify)
    6 national hero: other countries (specify)
    7 member of family (specify)
    8 other, including combinations (specify)
    9 unknown

| 64 | ____ | Attitude Toward Human Beings _____ |

    0 negative
    2 fluctuating
    3 positive
    5 dualistic
    8 other (specify)
    9 unknown

| 65 | ____ | Attitude Toward Own Country _____ |

    0 negative
    2 fluctuating
    3 positive
    8 other (specify)
    9 unknown

| 66 | ____ | Attitude Toward International Society _____ |

    0 negative
    2 fluctuating
    3 positive
    5 dualistic
    8 other (specify)
    9 unknown

| 67 | ____ | Religious Group _____ |

    0 main group of country
    1 minority group of country
    9 unknown

| 68 | ____ | Marital Status _____ |

    0 not married
    1 married
    9 unknown

Column(s)	Code	Variables and Response Values
69	___	Children _____

        0 none
        1 one
        2 two
        3 three
        4 four
        5 five
        6 six
        7 seven or more
        9 unknown

| 70 | ___ | National Birth _____ |

        0 domestic
        1 foreign
        9 unknown

| 71 | ___ | Region _____ |

        0 Africa
        1 Asia and the Middle East
        2 Latin America
        3 Europe and North America
        4 other (specify)
        9 unknown

Bibliography

Agawa, Hiroyuki. 1979. *The Reluctant Admiral: Yamamoto and the Imperial Navy.* Trans. by John Bester. Tokyo: Kodansha International.
Aldington, Richard. 1943. *The Duke: Being an Account of the Life and Achievements of Arthur Wellesley, 1st Duke of Wellington.* New York: Viking Press.
Ambrose, Stephen E. 1983. *Eisenhower: Soldier, General of the Army, President Elect.* New York: Simon & Schuster.
Ambrose, Stephen E. 1984. *Eisenhower: The President.* New York: Simon & Schuster.
Anderson, Nancy Scott and Dwight Anderson. 1988. *The Generals: Ulysses S. Grant and Robert E. Lee.* New York: Knopf.
Angell, Hildegard. 1930. *Simón Bolívar: South American Liberator.* New York: Norton.
[Anonymous]. [Hope, Eva?] 1885. *General Gordon: The Christian Hero.* New York: Crowell.
Armstrong, H. C. 1932. *Grey Wolf: Mustafa Kemal: An Intimate Study of a Dictator.* London: Arthur Barker.
Army Times Editors. 1960. *The Yanks Are Coming: The Story of General John J. Pershing.* New York: Putnam's.
Army Times Editors. 1967. *Warriors: The Story of General George S. Patton.* New York: Putnam's.
Arnault, M. A. and C. L. F. Panckoucke. 1855. *Life and Campaigns of Napoleon Bonaparte.* Two volumes in one. Boston: Phillips, Sampson, & Co.
Arthur, George. 1920. *Life of Lord Kitchener*, 3 vols. London: Methuen.
Ashley, Maurice. 1956. *Marlborough.* New York: Macmillan.
Aston, George. 1929. *The Biography of the Late Marshal Foch.* New York: Macmillan.
Athearn, Robert G. 1956. *William Tecumseh Sherman and the Settlement of the West.* Norman: University of Oklahoma Press.
Auty, Phyllis. 1970. *Tito: A Biography.* New York: McGraw-Hill.

Ayling, Keith. 1945. *Old Leatherface of the Flying Tigers: The Story of General Chennault*. Indianapolis: Bobbs-Merrill.
Barnett, Correlli. 1964. *The Swordbearers: Supreme Command in the First World War*. New York: Morrow.
Barnett, Correlli. 1974. *Marlborough*. London: Methuen.
Barnett, Correlli. 1978. *Bonaparte*. London: George Allen & Unwin.
Bass, Bernard M. and Dana L. Farrow. 1977. "Quantitative Analysis of Biographies of Political Figures." *The Journal of Psychology*, 97:281–96.
Beaumont, Roger A. 1974. *Military Elites*. Indianapolis: Bobbs-Merrill.
Bellamy, Francis R. 1951. *The Private Life of George Washington*. New York: Crowell.
Ben Shaul, Moshe, ed. 1968. *Generals of Israel*. Trans. by I. Hanoch. Tel Aviv: Hadar Publishing.
Bennett, Geoffrey. 1972. *Nelson the Commander*. London: B. T. Batsford.
Bernier, Oliver. 1983. *Lafayette: Hero of Two Worlds*. New York: Dutton.
Bevan, Bryan. 1975. *Marlborough the Man: A Biography of John Churchill, First Duke of Marlborough*. London: Robert Hale.
Blumenson, Martin. 1972–1974. *The Patton Papers*, 2 vols. Boston: Houghton Mifflin.
Blumenson, Martin and James L. Stokesbury. 1975. *Masters of the Art of Command*. Boston: Houghton Mifflin.
Bonaparte, Napoleon. 1856. *The Confidential Correspondence of Napoleon Bonaparte*, 2 vols. New York: Appleton.
Bonaparte, Napoleon. [1769–1821]. 1910. *The Corsican: A Diary of Napoleon's Life in His Own Words*. Trans. and ed. by R. M. Johnston. Boston: Houghton Mifflin.
Boulger, Demetrius C. 1911. *The Life of Gordon*. London: Fisher Unwin.
Bowman, S. M. and R. B. Irwin. 1865. *Sherman and His Campaigns: A Military Biography*. New York: Charles B. Richardson.
Brendon, Piers. 1986. *Ike: His Life and Times*. New York: Harper & Row.
Brock, Ray. 1954. *Ghost on Horseback: The Incredible Atatürk*. New York: Duel, Sloan and Pearce.
Brooks, William E. 1932. *Lee of Virginia: A Biography*. Indianapolis: Bobbs-Merrill.
Browne, Courtney. 1967. *Tojo: The Last Banzai*. New York: Holt, Rinehart & Winston.
Bryant, Arthur. 1971. *The Great Duke [Wellington]*. London: Collins.
Buck, James H. and Lawrence Korb, eds. 1981. *Military Leadership*. Beverly Hills, CA: Sage.
Butow, Robert J. C. 1961. *Tojo and the Coming of the War*. Princeton, NJ: Princeton University Press.
Byrd, Martha. 1987. *Chennault: Giving Wings to the Tiger*. Tuscaloosa: University of Alabama Press.
Callcott, Wilfrid Hardy. 1964. *Santa Anna: The Story of an Enigma Who Once Was Mexico*. Hamden, CT: Archon Books.
Carter, John H. 1952. "Military Leadership." *Military Review*, 32:14–18.
Carver, Michael. 1976. *The War Lords: Military Commanders of the Twentieth Century*. Boston: Little, Brown.

Castelot, André. 1971. *Napoleon*. Trans. by Guy Daniels. New York: Harper & Row.
Catherine II. N.d. [1771?] *The Memoirs of Catherine the Great*. Trans. from the French by Moura Budberg. Ed. by Dominique Maroger. New York: Macmillan.
Chalfont, Alun. 1976. *Montgomery of Alamein*. New York: Atheneum.
Chaney, Otto Preston, Jr. 1971. *Zhukov*. Norman: University of Oklahoma Press.
Chennault, Anna. 1963. *Chennault and the Flying Tigers*. New York: Paul S. Eriksson.
Chennault, Claire Lee. 1949. *Way of a Fighter: The Memoirs of Claire Lee Chennault*. Ed. by Robert Hotz. New York: Putnam's.
Chidsey, Donald Barr. 1930. *Marlborough: The Portrait of a Conqueror*. London: John Murray.
Churchill, Winston S. 1933–1938. *Marlborough: His Life and Times*, 6 vols. New York: Scribner's.
Clausewitz, Carl von. [1832]. 1968. *On War*. Ed. with an Introduction by Anatol Rapoport. London: Penguin Books.
Clausewitz, Carl von. [1832]. 1984. *On War*. Trans. and ed. by Michael Howard and Peter Paret. Introductory essays by Peter Paret, Michael Howard, and Bernard Brodie. Princeton, NJ: Princeton University Press.
Connelly, Owen. 1972. *The Epoch of Napoleon*. New York: Holt, Rinehart & Winston.
Connelly, Owen. 1987. *Blundering to Glory: Napoleon's Military Campaigns*. Wilmington, DE: Scholarly Resources, Inc.
Cook, Don. 1983. *Charles de Gaulle: A Biography*. New York: Putnam's.
Cooper, Leonard. 1963. *The Age of Wellington*. New York: Dodd, Mead.
Cornwall, James M. 1967. *Napoleon*. London: William Clowes & Sons.
Cronin, Thomas E. 1984. "Thinking About Leadership." In Robert L. Taylor and William E. Rosebach, eds. (1984), *Military Leadership: In Pursuit of Excellence*. Boulder, CO: Westview Press.
Cronin, Vincent. 1972. *Napoleon Bonaparte: An Intimate Biography*. New York: Morrow.
Crozier, Brian. 1967. *Franco: A Biographical History*. London: Eyre & Spottiswoode.
Crozier, Brian. 1973a. *De Gaulle: The Warrior*. London: Methuen.
Crozier, Brian. 1973b. *De Gaulle: The Statesman*. London: Methuen.
Cunliffe, Marcus. 1958. *George Washington: Man and Monument*. Boston: Little, Brown.
Curtis, James C. 1976. *Andrew Jackson and the Search for Vindication*. Boston: Little, Brown.
Dayan, Moshe. 1976. *Moshe Dayan: Story of My Life*. New York: Morrow.
Dayan, Yaël. 1985. *My Father, His Daughter*. New York: Farrar, Straus & Giroux.
Dediger, Vladimir. 1953. *Tito*. New York: Simon & Schuster.
De Gaulle, Charles. 1960. *The Edge of the Sword*. Trans. by Gerard Hopkins. New York: Criterion Books.
De Koven, Reginald. 1913. *The Life and Letters of John Paul Jones*, 2 vols. New York: Scribner's.
Del río, Daniel A. 1965. *Simón Bolívar*. N.p.: The Bolivarian Society of the United States.

DeLuca, Anthony R. 1983. *Personality, Power, and Politics.* Cambridge, MA: Schenkman.
Denikin, Anton I. [1953]. 1975. *The Career of a Tsarist Officer: Memoirs, 1872–1916.* An annotated translation from the Russian by Margaret Patoski. Minneapolis: University of Minnesota Press.
Dior Palen, Andreas. 1964. *Hindenburg and the Weimar Republic.* Princeton, NJ: Princeton University Press.
Dixon, Norman. 1976. *On the Psychology of Military Incompetence.* New York: Basic Books.
Douglas-Home, Charles. 1973. *Rommel.* London: Weidenfeld & Nicolson.
Dowdey, Clifford, 1965. *Lee.* Boston: Little, Brown.
Driault, Édouard. 1929. *The True Visage of Napoleon.* Trans. by W. Savage. Paris: Éditions Albert Morancé.
Dupuy, Trevor N., Curt Johnson, and David L. Bongard. 1992. *The Harper Encyclopedia of Military Biography.* New York: Harper Collins.
Earle, Edward M., ed. 1944. *Makers of Modern Strategy: Military Thought from Machiavelli to Hitler.* Princeton, NJ: Princeton University Press.
Edinger, George and E. J. C. Neep. 1931. *Nelson: The Life of Horatio Nelson.* New York: Jonathan Cape & Harrison Smith.
Elton, Lord. 1954. *General Gordon.* London: Collins.
Escher, Reginald. 1921. *The Tragedy of Lord Kitchener.* London: John Murray.
Essame, H. 1974. *Patton: A Study in Command.* New York: Scribner's.
Farago, Ladislas. 1964. *Patton: Ordeal and Triumph.* New York: Astor-Honor.
Faure, Élie. 1924. *Napoleon.* Trans. by Jeffrey E. Jeffrey. New York: Knopf.
Ferrell, Robert N. 1966. *George C. Marshall.* New York: Cooper Square.
Fisher, H. A. L. 1967. *Napoleon.* Oxford: Oxford University Press.
Flanner, Janet. 1944. *Pétain: The Old Man of France.* New York: Simon & Schuster.
Flexner, James T. 1965–1972. *George Washington,* 4 vols. Boston: Little, Brown.
Foch, Ferdinand. 1931. *The Memoirs of Marshal Foch.* Trans. by T. Bentley Mott. New York: Doubleday, Doran.
Fortescue, John. 1925. *Wellington.* New York: Dodd, Mead.
Freeman, Douglas Southall. 1934–1935. *R. E. Lee: A Biography,* 4 vols. New York: Scribner's.
Freeman, Douglas Southall. 1948–1957. *George Washington,* 7 vols. New York: Scribner's.
Freeman, Douglas Southall. 1968. *Washington: An abridgement in one volume by Richard Harwell of the seven-volume GEORGE WASHINGTON.* New York: Scribner's.
Frye, William. 1947. *Marshall: Citizen Soldier.* Indianapolis: Bobbs-Merrill.
Fuller, J. F. C. 1933. *Grant and Lee: A Study in Personality and Generalship.* London: Eyre & Spottiswoode.
Garrett, Richard. 1974. *General Gordon.* London: Arthur Barker.
Gibb, M. A. 1938. *The Lord General: A Life of Thomas Fairfax.* London: Lindsay Drummond.
Glover, Michael. 1968. *Wellington as Military Commander.* London: Batsford.
Glover, Michael. 1976. *General Burgoyne in Canada and America: Scapegoat for a System.* London: Gordon & Cremonesi.

Goebel, Dorothy Burne and Julius Goebel, Jr. 1945. *Generals in the White House.* New York: Doubleday, Doran.
Goodspeed, D. J. 1966. *Ludendorff: Genius of World War I.* Boston: Houghton Mifflin.
Gottschalk, Louis. 1942. *Lafayette and the Close of the American Revolution.* Chicago: University of Chicago Press.
Gottschalk, Louis and Margaret Maddox. 1973. *Lafayette in the French Revolution.* Chicago: University of Chicago Press.
Grenfell, Russell. 1950. *Nelson the Sailor.* New York: Macmillan.
Griffiths, Richard. 1970. *Marshal Pétain.* London: Constable.
Guedalla, Phillip. N.d. (1923?) *Men of War.* London: Hodder & Stoughton.
Guedalla, Phillip. 1943. *The Two Marshals: Bazaine, Pétain.* New York: Reynal & Hitchcock.
Gunther, John. 1951. *The Riddle of MacArthur.* New York: Harper & Bros.
Halsey, William F., Jr. and J. Bryan III. 1947. *Admiral Halsey's Story.* New York: McGraw-Hill.
Hamilton, Nigel. 1981. *Monty: The Making of a General, 1887–1942.* London: Hamish Hamilton.
Hamilton, Nigel. 1983. *Monty: Master of the Battlefield, 1942–1944.* London: Hamish Hamilton.
Hamilton, Nigel. 1987. *Monty: Final Years of the Field Marshal, 1944–1976.* New York: McGraw-Hill.
Handel, Michael I., ed. 1986. *Clausewitz and Modern Strategy.* London: Frank Cass.
Hanighen, Frank C. 1934. *Santa Anna: The Napoleon of the West.* New York: Coward McCann.
Hanson, Lawrence and Elizabeth Hanson. 1954. *Chinese Gordon: The Story of a Hero.* New York: Funk & Wagnalls.
Hargrove, Richard G. 1983. *General John Burgoyne.* Newark, NJ: University of Delaware Press.
Hatch, Alden. 1960. *The de Gaulle Nobody Knows: An Intimate Biography of Charles de Gaulle.* New York: Hawthorn Books.
Hayami, Yujiro, Vernon W. Ruttan, and Herman M. Southworth, eds. 1979. *Agricultural Growth in Japan, Taiwan, and the Philippines.* Honolulu: University of Hawaii Press.
Herold, J. Christopher, ed. and trans. 1955. *The Mind of Napoleon: A Selection From His Written and Spoken Words.* New York: Columbia University Press.
Hibbert, Christopher. 1966. *Garibaldi and His Enemies: The Clash of Arms and Personalities in the Making of Italy.* Boston: Little, Brown.
Hills, George. 1967. *Franco: The Man and His Nation.* New York: Macmillan.
Hindenburg, Marshal [Paul] von. 1920. *Out of My Life.* Trans. by F. A. Holt. London: Cassell.
Holmes, Richard. 1989. *Acts of War: The Behavior of Men in Battle.* New York: Free Press.
Hough, Richard. 1981. *Mountbatten.* New York: Random House.
Howard, Michael. 1983. *Clausewitz.* Oxford: Oxford University Press.
Howson, Gerald. 1979. *Burgoyne of Saratoga: A Biography.* New York: Times Books.
Hoyt, Edwin P. 1990. *Yamamoto: The Man Who Planned Pearl Harbor.* New York: McGraw-Hill.

Hunt, Frazier. 1954. *The Untold Story of Douglas MacArthur.* New York: Devin-Adair.
Hutt, Maurice, ed. 1972. *Napoleon.* Englewood Cliffs, NJ: Prentice-Hall.
Iremonger, Lucille. 1970. *The Fiery Chariot: A Study of the British Prime Ministers and the Search for Love.* London: Secker & Warburg.
Irving, David. 1977. *The Trail of the Fox.* New York: Dutton.
Jackson, Stuart W. 1930. *Lafayette: A Bibliography.* New York: Burt Franklin.
James, D. Clayton. 1970–1985. *The Years of MacArthur*, 3 vols. Boston: Houghton Mifflin.
James, Marquis. 1938. *The Life of Andrew Jackson.* Indianapolis: Bobbs-Merrill.
Janowitz, Morris. 1960, 1971. *The Professional Soldier: A Social and Political Portrait.* New York: Free Press.
Jenkins, William O. 1947. "A Review of Leadership Studies with Particular Reference to Military Problems." *Psychological Bulletin*, 44:54–79.
Jones, J. William. 1874. *Personal Reminiscences, Anecdotes, and Letters of Gen. Robert E. Lee.* New York: Appleton.
Jones, Oakah L., Jr. 1968. *Santa Anna.* New York: Twayne Publishers.
Journal of Applied Social Psychology. 1986. 16:461–575. Issue devoted to military psychology.
Kazancigil, Ali and Ergun Ozbudun, eds. 1981. *Atatürk: Founder of a Modern State.* Hamden, CT: Archon Books.
Keegan, John. 1978. *The Face of Battle.* New York: Viking Penguin.
Keegan, John. 1988. *The Masks of Command.* New York: Viking Penguin.
Keegan, John and Andrew Wheatcroft. 1976. *Who's Who in Military History: From 1453 to the Present Day.* London: Weidenfeld and Nicolson.
Kellett, Anthony. 1982. *Combat Motivation: The Behavior of Soldiers in Battle.* Boston: Kluwer-Nijhoff.
Kenez, Peter. 1971. *Civil War in South Russia, 1918.* Berkeley: University of California Press.
Kessel, Eberhard. 1957. *Moltke.* Stuttgart: K. F. Koehler Verlag.
Kinross, Lord. 1965. *Atatürk: A Biography of Mustafa Kemal, Father of Modern Turkey.* New York: Morrow.
Kircheisen, Friedrich M. 1931. *Nelson: Man and Admiral.* Trans. By Frederick Collins. New York: Duffield & Green.
Kircheisen, Friedrich M. 1932. *Napoleon.* Trans. by Henry St. Lawrence. New York: Harcourt, Brace & Co.
Klecka, William R. 1980. *Discriminant Analysis.* Beverly Hills, CA: Sage.
Kübler, Theodore. 1912. *General ("Chinese") Gordon: The Christian Hero.* Trans. by George P. Upton. Chicago: A. C. McClurg.
Lacouture, Jean. 1990. *De Gaulle: The Rebel, 1890–1944.* Trans. by Patrick O'Brian. New York: Norton.
Lacouture, Jean. 1991. *De Gaulle: The Ruler, 1945–1970.* Trans. by Alan Sheridan. New York: Norton.
Laffin, John. 1966. *Links of Leadership: Thirty Centuries of Command.* London: Harrap.
Lankin, Doris. 1968. "Moshe Dayan." In Moshe Ben Shaul, ed. (1968), *Generals of Israel.* Tel Aviv: Hadar Publishing.

Larg, David. [1934]. 1970. *Giuseppe Garibaldi: A Biography*. Port Washington, NY: Kennikat Press.
Laughlin, Clara E. 1919. *Foch the Man*. London: Revell.
Ledwidge, Bernard. 1982. *De Gaulle*. London: Weidenfeld & Nicolson.
Lee, Robert E., Jr. [1904]. 1960. *My Father, General Lee*. New York: Doubleday.
Lehovich, Dimitry V. 1974. *White Against Red: The Life of General Anton Denikin*. New York: Norton.
Lewin, Ronald. 1968. *Rommel as Military Commander*. London: Batsford.
Lewis, Lloyd. 1932. *Sherman: Fighting Prophet*. New York: Harcourt, Brace & Co.
Lewis, Paul. 1973. *The Man Who Lost America: A Biography of Gentleman Johnny Burgoyne*. New York: Dial Press.
Liddell-Hart, B. H. 1928. *Reputations: Ten Years After*. Boston: Little, Brown.
Liddell-Hart, B. H. 1931. *Foch: The Man of Orleans*. London: Eyre & Spottiswoode.
Liddell-Hart, B. H. [1958]. 1978. *Sherman: Soldier, Realist, American*. Westport, CT: Greenwood Press.
Lloyd, Alan. 1969. *Franco*. New York: Doubleday.
Long, A. L. 1887. *Memoirs of Robert E. Lee: His Military and Personal History*. New York: J. M. Soddart.
Long, Gavin. 1969. *MacArthur as Military Commander*. London: Batsford.
Longford, Elizabeth. 1969. *Wellington: The Years of the Sword*. New York: Harper & Row.
Longford, Elizabeth. 1972. *Wellington: Pillar of State*. London: Weidenfeld & Nicolson.
Lorenz, Lincoln. 1943. *John Paul Jones: Fighter for Freedom and Glory*. Annapolis, MD: Naval Institute Press.
Loth, David. 1951. *The People's General: The Personal Story of Lafayette*. New York: Scribner's.
Lottman, Herbert R. 1985. *Pétain: Hero or Traitor*. New York: Morrow.
Luckett, Richard. 1971. *The White Generals: An Account of the White Movement and the Russian Civil War*. London: Longman.
Ludendorff, Erich von. 1919. *Ludendorff's Own Story*, 2 vols. New York: Harper & Bros.
Ludendorff, Erich von. N.d. [1919]. *My War Memories, 1914–1918*. London: Hutchinson.
Ludwig, Emil. 1935. *Hindenburg*. Trans. by Eden and Cedar Paul. Philadelphia: John C. Winston.
Ludwig, Emil. 1942. *Bolívar: The Life of an Idealist*. New York: Alliance Book Corp.
Ludwig, Emil. [1924]. 1953. *Napoleon*. Trans. by Eden and Cedar Paul. New York: Modern Library.
Lyon, Peter. 1974. *Eisenhower: Portrait of a Hero*. Boston: Little, Brown.
MacArthur, Douglas. 1964. *Reminiscences*. New York: McGraw-Hill.
MacArthur, Douglas. 1965. *A Soldier Speaks*. New York: Praeger.
Machiavelli, Niccolò. [1521]. 1965. *The Art of War*. A revised edition of the Ellis Farneworth translation with an Introduction by Neal Wood. Indianapolis: Bobbs-Merrill.
Mackenzie, Alexander S. 1846. *The Life of John Paul Jones*, 2 vols. New York: Harper & Bros.
Magnus, Philip. 1959. *Kitchener: Portrait of an Imperialist*. New York: Dutton.

Magnusson, David, ed. 1981. *Toward a Psychology of Situations: An Interactional Perspective*. Hillsdale, NJ: Erlbaum.
Mahan, Alfred Thayer. 1890. *The Influence of Seapower Upon History, 1660–1783*. Boston: Little, Brown.
Mahan, Alfred Thayer. 1897. *The Life of Nelson: The Embodiment of the Seapower of Great Britain*, 2 vols. London: Sampson Low, Marston & Co.
Manchester, William. 1978. *American Caesar: Douglas MacArthur, 1880–1964*. Boston: Little, Brown.
Marquis Who's Who. 1975. *Who Was Who in American History—The Military*. Chicago: Marquis Who's Who.
Marshall, George C. 1976. *Memoirs of My Services in the World War, 1917–1918*. Boston: Houghton Mifflin.
Marshall-Cornwall, James. 1967. *Napoleon as Military Commander*. London: Batsford.
Marshall-Cornwall, James. 1972. *Foch as Military Commander*. London: Batsford.
Martell, Paul and Grace P. Hayes, eds. 1974. *World Military Leaders*. New York: R. R. Bowker.
Marszalek, John F. 1993. *Sherman: A Soldier's Passion for Order*. New York: Free Press.
Masur, Gerhard. 1948. *Simón Bolívar*. Albuquerque: University of New Mexico Press.
Matloff, Maurice and Stanley M. Ulanoff. 1985. *American Wars and Heroes: Revolutionary War Through Vietnam*. New York: Arco Publishing.
McCabe, James D., Jr. 1870. *Life and Campaigns of General Robert E. Lee*. Atlanta, GA: National Publishing Co.
McCracken, Harold. 1931. *Pershing: The Story of a Great Soldier*. New York: Brewer & Warren.
Meier, Norman C. 1943. *Military Psychology*. New York: Harper & Bros.
Mellenthin, F. W. von. 1977. *German Generals of World War II*. Norman: University of Oklahoma Press.
Mellor, William B. 1946. *Patton: Fighting Man*. New York: Putnam's.
Merrill, James M. 1971. *William Tecumseh Sherman*. Chicago: Rand McNally.
Merrill, James M. 1976. *A Sailor's Admiral: A Biography of William F. Halsey*. New York: Crowell.
Miller, Arthur H. 1920. *[Military] Leadership*. New York: Putnam's.
Mitcham, Samuel W., Jr. 1984. *Triumphant Fox: Erwin Rommel and the Rise of Afrika Korp*. New York: Stein & Day.
Mitchell, B. R. 1975. *European Historical Statistics, 1750–1970*. London: Macmillan.
Mitchell, B. R. 1988. *British Historical Statistics*. Cambridge: Cambridge University Press.
Montgomery, Bernard Law. 1958. *The Memoirs of Field Marshal the Viscount Montgomery of Alamein*. Cleveland, OH: World Publishing.
Montgomery, Bernard Law. 1961. *The Path to Leadership*. London: Collins.
Moorehead, Alan. 1946. *Montgomery: A Biography*. New York: Coward-McCann.
Morison, Samuel Eliot. 1959. *John Paul Jones: A Sailor's Biography*. Boston: Little, Brown.
Mosley, Leonard. 1982. *Marshall: Hero for Our Time*. New York: Hearst Books.
Mylander, Maureen. 1974. *The Generals*. New York: Dial Press.

Nutting, Anthony. 1966. *Gordon of Khartoum: Martyr and Misfit.* New York: Clarkson N. Potter.
O'Connor, Richard. 1961. *Black Jack Pershing.* New York: Doubleday.
O'Connor, Sandra Day. 1982. Quoted in *New York Times.* July 11.
Oman, Carola. 1947. *Nelson.* London: Hodder & Stoughton.
Otis, James. 1900. *The Life of John Paul Jones.* New York: A. L. Burt.
Paret, Peter. 1976. *Clausewitz and the State.* New York: Oxford University Press.
Parkinson, Roger. 1970. *Clausewitz: A Biography.* London: Wayland.
Parkinson, Roger. 1978. *Tormented Warriors: Ludendorff and the Supreme Command.* New York: Stein & Day.
Patton, George S. 1947. *War as I Knew It.* Annotated by Paul D. Harkins. Boston: Houghton Mifflin.
Pavlowitch, Steven K. 1992. *Tito: Yugoslavia's Great Dictator: A Reassessment.* Columbus: Ohio State University Press.
Payne, Robert. 1951. *The Marshall Story: A Biography of General George C. Marshall.* New York: Prentice-Hall.
Pershing, John J. 1918. *General Pershing's Own Story.* Philadelphia: George Barrie Sons.
Pershing, John J. 1931. *My Experiences in the World War,* 2 vols. New York: Frederick A. Stokes.
Pocock, Tom. 1988. *Horatio Nelson.* New York: Knopf.
Pogue, Forrest C. 1963. *George C. Marshall: Education of a General, 1889–1939.* New York: Viking Press.
Potter, E. B. 1976. *Nimitz.* Annapolis, MD: Naval Institute Press.
Potter, E. B. 1985. *Bull Halsey.* Annapolis, MD: Naval Institute Press.
Puleston, W. D. 1939. *Mahan: The Life and Work of Captain Alfred Thayer Mahan, U.S.N.* New Haven, CT: Yale University Press.
Puryear, Edgar F., Jr. 1971. *Nineteen Stars.* Washington, DC: Coiner Publications.
Rawls, Walton, ed. 1985. *Great Civil War Heroes and Their Battles.* New York: Abbeville Press.
Recouly, Raymond. 1929. *Foch: My Conversations with the Marshal.* Trans. by Joyce Davis. New York: Appleton.
Rejai, Mostafa and Kay Phillips. 1979. *Leaders of Revolution.* Beverly Hills, CA: Sage.
Rejai, Mostafa and Kay Phillips. 1983. *World Revolutionary Leaders.* New Brunswick, NJ: Rutgers University Press.
Rejai, Mostafa and Kay Phillips. 1988. *Loyalists and Revolutionaries: Political Leaders Compared.* New York: Praeger.
Rejai, Mostafa and Kay Phillips, with Warren L. Mason. 1993. *Demythologizing an Elite: American Presidents in Empirical, Comparative, and Historical Perspective.* Westport, CT: Praeger.
Remini, Robert V. 1977. *Andrew Jackson and the Course of American Empire, 1767–1821.* New York: Harper & Row.
Remini, Robert V. 1981. *Andrew Jackson and the Course of American Freedom, 1822–1832.* New York: Harper & Row.
Remini, Robert V. 1984. *Andrew Jackson and the Course of American Democracy, 1831–1835.* New York: Harper & Row.

Roberts, Lord. 1895. *The Rise of Wellington*, 2nd ed. London: Sampson Low, Marston & Co.

Robinson, Donald. 1970. *The 100 Most Important People in the World Today*. New York: Putnam's.

Rommel, Erwin. [1937, 1950]. 1953. *The Rommel Papers*. Ed. by B. H. Liddell-Hart, Lucie-Maria Rommel, Manfred Rommel, and Fritz Bayerlein. Trans. by Paul Findlay. New York: Harcourt, Brace.

Rose, J. Holland. 1912. *The Personality of Napoleon*. New York: Putnam's.

Sampson, Jack. 1987. *Chennault*. New York: Doubleday.

Santa Anna, Antonio Lopéz de. [1874]. 1967. *The Eagle: The Autobiography of Santa Anna*. Ed. by Ann Fears Crawford. Austin, TX: The Pemberton Press.

Sarkesian, Sam C. 1981. "A Personal Perspective." In James H. Buck and Lawrence Korb, eds. (1981), *Military Leadership*. Beverly Hills, CA: Sage.

SAS Institute. 1982. *SAS User's Guide*. Cary, NC: SAS Institute.

Schoenbrun, David. 1966. *The Three Lives of Charles de Gaulle*. New York: Atheneum.

Seager, Robert II. 1977. *Alfred Thayer Mahan: The Man and His Letters*. Annapolis, MD: Naval Institute Press.

Semmes, Harry H. 1955. *Portrait of Patton*. New York: Appleton-Century-Crofts.

Sharp, Paul S. 1981. *The East European and Soviet Data Handbook: Political, Social, and Developmental Indicators, 1945–1975*. Stanford: Hoover Institution Press; New York: Columbia University Press.

Sherman, William T. 1875. *Memoirs*, 2 vols. New York: Appleton.

Smith, Denis Mack, ed. 1969. *Great Lives Observed: Garibaldi*. Englewood Cliffs, NJ: Prentice-Hall.

Smythe, Donald. 1973. *Guerrilla Warrior: The Early Life of John J. Pershing*. New York: Scribner's.

Smythe, Donald. 1986. *Pershing: General of the Armies*. Bloomington: Indiana University Press.

Soloveytchik, George. 1947. *Potemkin: Soldier, Statesman, Lover, and Consort of Catherine of Russia*. New York: Norton.

Southey, Robert. 1962. *Life of Nelson*. London: J. M. Dent.

Spears, Edward, 1966. *Two Men Who Saved France: Pétain and de Gaulle*. New York: Stein & Day.

Spiller, Roger J. and Joseph G. Dawson III, eds. 1989. *American Military Leaders*. New York: Praeger.

SPSS^x User's Guide, 3d ed. 1988. Chicago: SPSS, Inc.

Stokesbury, James L. 1981. "Leadership as an Art." In James H. Buck and Lawrence Korb, eds. (1981), *Military Leadership*. Beverly Hills, CA: Sage.

Stoler, Mark A. 1989. *George C. Marshall: Soldier-Statesman of the American Century*. Boston: Twayne.

Sun Tzu. [4th century B.C.]. 1971. *The Art of War*. Trans. and with an Introduction by Samuel B. Griffith. New York: Oxford University Press.

Taeuber, Irene B. 1958. *The Population of Japan*. Princeton, NJ: Princeton University Press.

Taylor, Robert L. and William E. Rosenbach, eds. 1984. *Military Leadership: In Pursuit of Excellence*. Boulder, CO: Westview Press.

Terraine, John. 1968. *The Life and Times of Lord Mountbatten*. London: Hutchinson.

Teveth, Shabtai. 1972. *Moshe Dayan.* London: Weidenfeld & Nicolson.
Tongas, Gérard. [1937]. 1939. *Atatürk and the True Nature of Modern Turkey.* Trans. from the French by F. F. Rynd. London: Luzak.
Tournoux, Jean-Raymond. 1966. *Sons of France: Pétain and de Gaulle.* Trans. by Oliver Coburn. New York: Viking.
Trench, Charles C. 1978. *Charley Gordon: An Eminent Victorian Reassessed.* London: Allen Lane.
Trythall, J. W. D. 1970. *Franco: A Biography.* London: Rupert Hart-Davis.
Vagts, Alfred. 1937. *A History of Militarism.* New York: Norton.
Vandiver, Frank E. 1977. *Black Jack: The Life and Times of John J. Pershing,* 2 vols. College Station, TX: Texas A&M University Press.
Van Fleet, David D. and Gary A. Hukl. 1986. *Military Leadership: An Organizational Behavior Perspective.* Greenwich, CT: JAI Press.
Viotti, Andrea. 1979. *Garibaldi: The Revolutionary and His Men.* Dorset, England: Blandford Press.
Volkan, Vamik D. and Norman Itzkowitz. 1984. *The Immortal Atatürk: A Psychobiography.* Chicago: University of Chicago Press.
Walters, John B. 1973. *Merchant of Terror: General Sherman and Total War.* Indianapolis: Bobbs-Merrill.
Ward, Geoffrey, C. 1992. "Douglas MacArthur: An American Soldier." *National Geographic* (March), 181:54–83.
Warner, Oliver. 1958. *Victory: The Life of Lord Nelson.* Boston: Little, Brown.
Warner, Phillip. 1985. *Kitchener: The Man Behind the Legend.* London: Hamish Hamilton.
Watson, Thomas E. 1902. *Napoleon: A Sketch of His Life, Character, Struggles, and Achievements.* New York: Macmillan.
Wellard, James. 1946. *General George S. Patton, Jr.: Man Under Mars.* New York: Dodd, Mead.
Westrate, Edwin V. 1936. *Those Fatal Generals.* New York: Knight Publications.
Wheeler-Bennett, John W. 1936. *Wooden Titan: Hindenburg in Twenty Years of German History, 1914–1934.* New York: Morrow.
Wilson, John. 1985. *Fairfax.* New York: Franklin Watts.
Wilson, Woodrow. [1890]. 1952. *Leaders of Men.* Ed. with an Introduction and Notes by T. H. Vail Mottes. Princeton, NJ: Princeton University Press.
Witty, Paul A. and Harvey C. Lehman. 1932. "Nervous Instability and Genius: Military and Political Leaders." *Journal of Social Psychology,* 3:212–33.
Wood, W. J. 1984. *Leaders and Battles: The Art of Military Leadership.* Novato, CA: Presidio Press.
Worcester, Donald E. 1977. *Bolívar.* Boston: Little, Brown.
Ybarra, T. R. 1929. *Bolívar: The Passionate Warrior.* New York: Ives Washburn.
Young, Desmond. 1950. *Rommel: The Desert Fox.* New York: Harper & Bros.
Zhukov, Georgi K. [1969]. 1971. *The Memoirs of Marshal Zhukov.* New York: Delacorte Press.
Ziegler, Philip. 1985. *Mountbatten: The Official Biography.* London: Collins.
Ziegler, Philip, ed. 1989. *From Shore to Shore: The Tour Diaries of Earl Mountbatten of Burma, 1953–1979.* London: Collins.

Index

Adams, John, 41–42, 45
Age, and achievement, 4–5, 14, 123
Agnosticism, 19, 21
Ambition, 53, 66, 70; lack of, 101
American Caesar (Manchester), 57
Arrest, 23, 99, 124
Atatürk, Mustafa Kemal (né Mustafa), 14, 112–14, 129, 131, 134
Atheism, 19
Attitude, 26, 28

Bereavement, 7
Birth order, 5, 18–19, 124
Birthplace, 14, 132–33
Bolívar, Simón, 14, 114–17, 129, 134
Bonaparte, Napoleon, 14, 77–80, 131, 132, 133; as hero, 83–84, 115; imperialism of, 125; marginality of, 129–30; turning points for, 77–78
Burgoyne, John, 18, 66–67, 130
Byrd, Martha, 62

Carter, Ann Hill, 48, 49
Catherine II (the Great), 46, 106–7
Chance, 8, 39, 135
Chennault, Claire Lee, 61–62
Chennault, Jessie Lee (mother), 61

Chennault, John Stonewall Jackson (father), 61
Chennault, Lottie Barnes (stepmother), 61
Churchill, John, 66
Churchill, Winston (father), 66
Clausewitz, Carl von, 89–91, 130
Clausewitz, Friedrich Gabriel von (father), 89
Communist Party, 118–19
Compensation, 50–51, 67
Compulsion, 7
Cosmopolitanism, 5–6, 23, 124
Crises, 134–35; types of, 8
Culture, urban, 5, 14, 123

Dayan, Dvorah (mother), 97–98
Dayan, Moshe, 14, 97–99, 129, 134
Dayan, Ruth Shwarz, 98–99
Dayan, Shmuel (father), 97–98
Death wish, 69, 72, 131
De Gaulle, Charles, 85–87, 130, 131
Denikin, Anton Ivanovich, 14, 107–9
Depression, 50, 52, 68, 90, 91, 130
Deprivation. See Love deprivation; Relative deprivation
Discriminant analysis, 31, 33

Dixon, Norman, 6
Duty, 49

Education, 5, 21, 124; lack of, 41–42, 104
Egotism, 7, 37; of Bolívar, 116; of de Gaulle, 87; of Lafayette, 84; of MacArthur, 57, 131; of Mahan, 51; of Napoleon, 130–31; of Patton, 131; of Santa Anna, 131; of Zhukov, 110
Eisenhower, Dwight D., 14, 62–63, 125, 135
Ethnicity, 5, 19, 123
Ewing, Thomas, 49–50, 133
Exile, 23, 124
Experiences, 23

Factor analysis, 31, 32–33
Fairfax, Thomas, 14, 65–66, 130
Family life, 5, 18–19, 74, 97–98, 133–34
Family tradition, 8, 38–39, 65
Fatalism, 70, 72
Father: occupation of, 5, 21, 124; relationship with, 54–55, 74, 98
Foch, Ferdinand, 14, 83–84, 131
Franco, Francisco, 111–12, 133
Franco, Nicólas (father), 111

Garibaldi, Giuseppe, 14, 16, 99–101, 131
Gordon, Charles George, 70–72, 131, 133
Gordon, Elizabeth Enderby (mother), 70
Gordon, Henry William (father), 70

Haganah, 99
Halsey, William F., Jr., 57–58, 132
Halsey, William F., Sr., 57–58
Hamilton, Emma, 69
Hero worship, 26, 28, 83–84
Hewes, Joseph, 46
Hindenburg, Paul von, 14, 92–93, 95, 130, 133
Hitler, Adolf, 95, 96
Honor, from promotion, 43
Human nature, view of, 6, 28, 124
Humboldt, Alexander von, 116

Identity crisis, 101
Ideology, 13; and age, 14; exposure to, 4–5, 123; type of, 6, 23, 26, 124
Illegitimacy, 18, 66–67
Imperialism: of Bonaparte, 125; of Mahan, 51, 125; of Tojo, 125; of Yamamoto, 125
Imprisonment, 23, 99, 124
Inferiority complex, 92, 130
Influence of Seapower Upon History, 1660–1783, The (Mahan), 51
International society, view of, 28, 124
Iremonger, Lucille, 7

Jackson, Andrew, 16, 47–48, 129, 130, 134
Janowitz, Morris, 6
Jones, John Paul (né John Paul, Jr.), 45–46, 133

Keegan, John, 7
Kitchener, Frances Ann (mother), 72, 73
Kitchener, Henry (father), 72–73
Kitchener, Horatio Herbert, 72–73
Kokutai, 103–4

Lafayette, Gilbert, 14, 80–83, 129, 130
Land, acquisition of, 44
Leader-subordinate relations, 6
Lee, Henry (father), 48–49
Lee, Robert E., 48–49, 133, 134
Lehman, Harvey C., 6
Louis of Battenberg (prince), 75–76
Love deprivation, 7, 37, 38; of Atatürk, 129; of Bolívar, 115, 129; of Chennault, 61; of Dayan, 129; of Fairfax, 65; of Jackson, 47, 48, 129; of Kitchener, 72, 73; of Lafayette, 80–81, 129; of Moltke, 91, 92; of Montgomery, 129; of Nelson, 67–68; of Pétain, 84; of Sherman, 50; of Washington, 41, 129
Luck. *See* Chance
Ludendorff, Erich von, 14, 94–95, 134
Ludwig, Emil, 93

MacArthur, Arthur, Jr., 56
MacArthur, Douglas, 55–57, 131, 133
Magnusson, David, *Toward a Psychology of Situations: An Interactional Perspective*, 4
Mahan, Alfred Thayer, 14, 51–52, 130, 131, 133; imperialism of, 125; *The Influence of Seapower Upon History, 1660–1783*, 51; on Nelson, 68
Mahan, Dennis Hart (father), 51
Manchester, William, *American Caesar*, 57
Marginality, 37, 38; of Bonaparte, 78–79, 129–30; of Burgoyne, 66–67, 130; of Clausewitz, 130; of de Gaulle, 130, 131; of Fairfax, 66, 130; of Hindenburg, 130; of Jackson, 130; of Lafayette, 130; of Mahan, 130; of Marshall, 53, 54, 130; of Moltke, 130; of Nelson, 68, 130; of Pershing, 130; types of, 129–30; of Washington, 130; of Wellesley, 130
Marshall, George C., 53–55, 130
Marshall, George, Sr., 53, 54
Marszalek, John F., 50
Maxwell, Mungo, 46
Merchant of Terror (Walters), 51
Military leaders: characteristics of, 4–6, 13–33; motivational variables in, 6–9, 37–119, 135–36; psychological variables in, 37–119, 125, 129–32; situational variables in, 37–119, 132–35; sociodemographic variables in, 4–6, 13–33, 123–24
Moltke, Friedrich von (father), 91–92
Moltke, Helmuth von, 14, 91–92, 129, 130
Moltke, Henriette von (mother), 92
Montgomery, Bernard Law, 73–75, 131
Montgomery, Henry (father), 73, 74–75
Montgomery, Maud (mother), 73, 74
Mother: devotion to, 49, 54, 57, 61, 73, 98; relationship with, 41, 43, 69–70, 74; support from, 92
Mountbatten, Louis, 75–76
Mylander, Maureen, 6

Mysticism, 131; of Gordon, 72; of Halsey, 58; of Montgomery, 75; of Napoleon, 80; of Nelson, 69; of Patton, 59; of Wellesley, 70. *See also* Religion

Narcissism, 37, 130–31
National development, as control variable, 13
Nationalism, 7, 28, 37, 38, 124, 125; of Atatürk, 113; of de Gaulle, 86–87; of Ludendorff, 94–95; of Napoleon, 78; of Potemkin, 106–7; of Santa Anna, 104–5; of Tito, 117, 118–19; of Tojo, 102–3; of Yamamoto, 102
Nationalist movements, 8
Nelson, Catherine (mother), 67
Nelson, Edmund (father), 67
Nelson, Horatio, 51, 67–69, 130, 131, 133
Nimitz, Chester W., 60–61, 135

Occupation, 124; of father, 5, 21, 124; and social class, 21
O'Connor, Sandra Day, 4
Organizations, membership in, 23, 99, 124

Pacifism, 62
Parochialism, 23
Patton, George S., 58–59, 131
Patriotism, 7, 86, 103
Pershing, John J., 52–53, 130, 135
Personality, dual, 61, 62. *See also* Psychology
Personal traits, 9
Pétain, Omer Venant (father), 84
Pétain, Philippe, 84–85, 133–34
Politics, 5, 6, 44–45, 48
Potemkin, Alexander (father), 106
Potemkin, Grigory Alexandrovich, 106–7
Power, 79
Psychology, 3; components of, 7–8; findings, 125, 129–32; variables of, 37–38, 39

Racism, 51, 52, 53, 103, 125
Relative deprivation, 7, 37, 38; of Churchill, 66, 125, 129; of Clausewitz, 89; of Deniken, 107, 108; of Eisenhower, 62–63, 125; of Jackson, 47, 48; of Jones, 45; of Marshall, 53; of Moltke, 91, 92, 129; of Nelson, 67–68; of Pershing, 52, 125; of Sherman, 50; of Tito, 117–18, 129; of Yamamoto, 101; of Zhukov, 109
Religion, 5, 123, 131; background versus orientation, 19, 21; and de Gaulle, 86; and Foch, 84; and Gordon, 70, 71–72; and Mahan, 51, 52; and Patton, 59; and vanity, 57. *See also* Mysticism
Risorgimento, 99–100
Rommel, Erwin, 95–96, 134

Santa Anna, Antonio Lopéz de, 104–6, 131, 134
Sartre, Jean Paul, 87
Schnarnhorst, Gerhard von, 90
Self–glorification, 105–6, 110
Sherman, Charles Robert (father), 49–50
Sherman, William Tecumseh, 49–51
Siblings, 5, 18, 124; relationship with, 54–55
Situation, 3, 124; types of, 8–9; variables in, 38–39, 132–35
Social class, 5, 13, 52, 92; changes in, 124; designation of, 14, 16, 18; and occupation, 21
Sociodemography, 33, 123–24
Stalin, Joseph, 110, 119
Suckling, Maurice, 67, 68, 133
Superstition, 58, 80, 131, 132

Talmud, 4
Time period: and age, 14; as control variable, 13; and family life, 19; and social class, 16
Tito, Josip Broz (né Josip Broz), 117–19, 129, 134–35

Tojo, Hideki, 102–4, 125
Toward a Psychology of Situations: An Interactional Perspective (Magnusson), 4
Tradition, family, 8, 38–39, 65
Tuchman, Barbara, 4

Ultranationalism, 37, 38; of Franco, 111, 112; of Tojo, 103–4
Upward social mobility, 16, 23
Urban culture, 5, 14, 123

Vanity, 7, 37, 38; of Bolívar, 116; of Burgoyne, 67; of de Gaulle, 87; of Foch, 84; of Lafayette, 83; of MacArthur, 57, 131; of Mahan, 51–52; of Montgomery, 75; of Napoleon, 130–31; of Patton, 131; and religion, 57; of Santa Anna, 104–5, 131

Walters, John B., *Merchant of Terror*, 51
War, passion for, 71
Washington, George, 41–45, 48, 80, 82, 134; marginality of, 130
Washington, Lawrence (brother), 41, 42, 43
Washington, Martha Custis, 42
Washington, Mary Ball (mother), 42
Wellesley, Anne (mother), 69
Wellesley, Arthur, 69–70, 130, 131, 134
Wellesley, Garrett (father), 69
Wheeler-Bennett, John W., *Wooden Titan*, 93
Witty, Paul A., 6
Wooden Titan (Wheeler-Bennett), 93

Yamamoto, Isoroku (né Takano), 101–2, 125, 133

Zhukov, Georgi Konstantinovich, 109–11

About the Authors

MOSTAFA REJAI is Distinguished Professor Emeritus at Miami University, Ohio, where he has also been the recipient of an Outstanding Teaching Award.

KAY PHILLIPS is Professor of Sociology and Anthropology at Miami University, Ohio.